How Our Brain Works

The Construction and Functionality of Your Brain Presented and Explained

Donald A. Millers

iUniverse, Inc.
New York   Bloomington

iUniverse books may be ordered through booksellers or by contacting:

iUniverse
1663 Liberty Drive
Bloomington, IN 47403
www.iuniverse.com
1-800-Authors (1-800-288-4677)

Because of the dynamic nature of the Internet, any Web addresses or links contained in this book may have changed since publication and may no longer be valid. The views expressed in this work are solely those of the author and do not necessarily reflect the views of the publisher, and the publisher hereby disclaims any responsibility for them.

ISBN: 978-1-4502-0877-2 (sc)
ISBN: 978-1-4502-0878-9 (ebook)
ISBN: 978-1-4502-2037-8 (dj)

Printed in the United States of America

iUniverse rev. date: 03/09/2010

For my wife Janna and my

daughters Traci and Samantha

Contents

List of Illustrations

Introduction

The attributes of the human brain that produce human intelligence are the focus of this book. This book is about the real you. Not the reflected image you see in the mirror, the shape of your face and body and the coloration of your skin, hair, and eyes. That's a machine. A set of moveable connected bones driven by muscles and covered with skin and hair. You control the machine that you see in the mirror. The real you is contained within your brain.

At this very moment, your brain is building neural representations of every object you currently perceive and a neural representation of the space around you. These perceptions of objects and space, contained in different areas of your brain, are brought together to build your virtual world. That neural virtual world is your reality. The most important object in your virtual world is you. Your neural representation of yourself is much larger and more complicated than what you see in the mirror.

In our conversations about the human brain, we will treat a very complex subject in a simple manner. The words simple and complex are terms that often describe our understanding of things rather than descriptions of the things themselves. Things we understand, we tend to describe as simple; things we do not understand, we tend to describe as complex. The brain is commonly referred to as the most complex structure in the known universe. This description is a reflection of our current lack of understanding as well as a tribute to the vast numbers of microscopic neuron cells, each with multiple thousands of interconnections to other neurons, that make up the brain.

Most of your neural system is inherited. The neural circuits that comprise this portion of you are genetically determined, are identical in all humans, and remain static throughout your entire life. A few neural components, the cerebellum, basal ganglia and cerebral cortex, contain large arrays of identical neural circuits that modify their wiring with learning. It is these modified by experience neural components that enable human intelligence.

I am going to assume that your familiarity with neurons and neural components is limited. This assumption creates an immediate problem. It is very difficult to discuss any aspect of human brain function without an understanding of neuron functionality and basic brain structure. An in depth discussion requires a great deal of prior knowledge. We will therefore divide our time together into two parts, an introductory journey through your brain followed by an in depth discussion and analysis.

We will begin each of our two explorations with a discussion of neurons in chapters one and four. Neurons are the basic elemental building material from which brains are constructed. It is necessary to understand what they do and perhaps as importantly, what they don't do. Neurons do not store data. Neurons do not perform computations. Performing computations on stored data are computer like functions. You cannot utilize neurons to build a computer. We are going to examine how neurons build a brain.

Neurons are complex living cells that perform one function, pattern detection. What neurons are, how they work, and what they do is the purview of chapter one. Chapter four examines the attributes of neurons that enable learning and the need for, and capability of, neurons to support synchronous activity.

Human intelligence is facilitated by a vast array of neurons performing various pattern detection tasks.

The neural components that make up your nervous system will be introduced in chapter two. Your neural structure will be compared to the more familiar architectural structure of a five-story building. The various components that comprise the floors of that building will be discussed in some detail. After we have examined and understood these floors, we will re-examine each of the various neural components within actual brain architecture in chapter five, exploring their functions and how they accomplish those functions.

The first floor of your brain building contains the elevator shaped spinal cord. Spinal cords have been around for over four hundred million years. Every spinal cord on earth today is just like yours. The second floor contains the brain stem, your reptilian brain. The brain stem is mostly a maintenance service center controlling temperature, blood pressure, heart rate and other functions necessary to keep the body alive and functioning. The second floor also contains a movement-learning center called the cerebellum. The cerebellum is where you store the patterns for "How to ride a bike".

There are two main rooms on the third floor that contain your chemistry lab, the hypothalamus, and your main switchboard, the thalamus. The hypothalamus is the master component of your limbic system. The limbic system controls behavior through the release of hormones. Quite specific, behaviorally correct, complicated behaviors are turned on and off by the limbic system's control of the chemical state of the body. Watch a pair of birds perform an intricate, beautiful mating dance or feed their young to witness this type of chemically controlled behavior.

An output control center called the basal ganglia completely occupies the fourth floor. The fourth floor controls your behavior. The fifth floor houses your executive level functions contained within your cerebral cortex. The storage of all the neural patterns that drive the higher-level neural capabilities that make you human are contained on the fifth floor. This is where memory patterns are recognized.

Memory recognition in the cerebral cortex is enabled by large pyramidal neurons that have an average of 60,000 inputs and 20,000 outputs. A pyramidal neuron monitors its 60,000 inputs and signals to 20,000 other neurons its level of pattern recognition. Morphing this description to be more human friendly, 60,000 inputs can be represented as a black and white TV with a 245 by 245 pixel array. Not great resolution but a 245 by 245 array can display quite complicated patterns.

Risking a giant anthropomorphic distortion of reality, the pyramidal neuron "watches" its black and white TV screen and signals to 20,000 other pyramidal neurons the level of pattern recognition it has for the current image. There are approximately fifteen billion pyramidal neurons in your cerebral cortex. The pattern complexity that can be represented by fifteen billion low-resolution black and white TV sets is staggering.

There are numerous examples in the animal kingdom of one, two and three story brains. Each of these fewer floor brain-building architectures represent complete functioning neural systems sustaining the animals they serve. Brains have evolved to add more floors and enlarge existing floors. All mammals have five floors and all of these floors are very similar in all mammals. It is the fifth floor cerebral cortex and interconnected neural components that have expanded in size to enable the mental capacity that makes you human.

The aspects of human intelligence we are trying to understand, memory, learning, intelligence, and behavior, will be introduced in chapter three. Each is presented and dissected. A highly speculative discussion of human brain evolution is also presented in chapter three. The human nervous system was not designed, it was grown. New structures that enabled new capabilities were added on top of existing neural systems through evolution. Each new evolutionary stage of neural development represents a complete, operational nervous system. Neural subsystems were not discarded and replaced by new designs. After around 500 million years of this process, human intelligence emerged.

The sixth and final chapter ties all of the neural components together into a complete architectural model of your brain. Based on this architectural model and the sum total of all our prior deliberations, explanations of memory, learning, and behavior are presented.

This book contains a great deal of speculation. This is the first of many such admonishments that will be sprinkled throughout the upcoming text. The sum total of all of those warnings pales in comparison to the actual level of speculation presented. A lens of caution should be applied to all that follows. The logical progression of data and analysis presented will hopefully justify all predictions of neurological function. You will be the final judge of that.

Who am I, besides being a person who is curious about his brain, I am an engineer. I have spent my entire career in the development of digital systems, both hardware and software. The frustrating experience of researching how my brain actually works became the genesis of this book. This is the book I wanted to read. It is important to note that I am not a scientist. I have not contributed to the body of knowledge that constitutes the current state of neuroscience. I am utilizing that body of knowledge in order to articulate what we do know and analyzing that body of knowledge in order to inform what we do not. These explanations of neurological functions seem to best fit the data available.

There are many questions about the brain that most of us find interesting. How does memory work? Is your entire life stored inside your head? What determines what you can remember? How do you learn? How do you control your behavior? What roll do your emotions play in the operation of your brain? How do you control the machine that you see when you look into a mirror? Your brain is a truly amazing organ. Lets begin the process of understanding its structure, organization, and capabilities.

Chapter 1

So You Want To Build A Brain You're Going To Need Some Neurons

Brain Architecture

The dictionary's definition of architecture, "The manner in which the elements (as of a design) are arranged or organized", applies equally as well to human brain architecture as it does to building architecture. An actual architectural design is constrained by the construction materials utilized. Mud and straw, fired bricks, and steel beams each allow the designer a very different degree of artistic freedom in the actual construction of a building. Skyscrapers cannot be built with bricks. The only construction material available for the building of brains is the neuron.

   The neuron is in many ways like a brick. Both bricks and neurons have very limited actual contact with others of their kind. Each is relatively isolated from the greater whole. Each performs a similar task as all the other bricks and neurons. Bricks and neurons require support from their contact neighbors in order to provide their function. A single brick or neuron has no utility. Each combines with others to produce a larger entity, the collection of all of those entities defines the whole. The whole is made up of a very large number of bricks or neurons, each performing a common task.

Our attempt to understand the human brain starts with an analysis of the brain's basic brick, the neuron. The human brain was not designed. It was grown over hundreds of millions of years of evolution. Be that as it may, the growth of the human brain to its present architectural design was constrained by the limitations of neurons. The neuron defines what a brain can be.

In this chapter you will learn what a neuron is, what it does, and how it combines with other neurons to form neural circuits. For our purposes we will define two distinct types of neural circuits, hardwired and learning neural circuits. The neural wiring diagram of much of the human nervous system is determined genetically and is not altered with experience. We will refer to this type of neural circuit as hardwired. This hardwired portion of your neural system tends to be very old and is similar in all vertebrates. Your body does not learn to regulate your body temperature and blood pressure. The neural circuits that have evolved to control these types of functions are fixed or, by our definition, hardwired. This is not a common distinction but we will find it useful in our analysis.

In the latest evolutionary additions to the brain, the neural wiring is more malleable or plastic in nature. Neurons grow, shrink or die depending on the demands made of them. It is this capability of neurons to change that makes memory and learning possible. Learning type neurons recognize previously encountered patterns of input. These neurons tend to have very large numbers of inputs and outputs.

The Neuron (The Only Component Available for Building Brains)

The neuron is a specialized cell with all of the cellular apparatus of a normal cell. What makes the neuron cell unique are its shape and signaling characteristics. Neurons exhibit a common, similar structure. (Ref. Fig. 1-1) The cell body serves as the focal point for one or more dendritic inputs and one axonal output. The dendritic input portion of the neuron collects signaling input from other neurons in a limited volume of space. The one axonal output transports the neuron's output signal to target areas and then divides to signal multiple neurons.

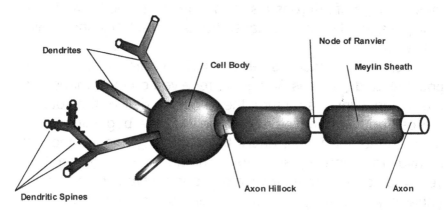

Figure 1-1: Neuron

Interconnections between neurons are called synapses and are not actually physical connections. Neurons communicate across small synaptic gaps that support all input and output intercellular communication. The synapses of learning type neurons are typically located on small input structures called dendritic spines.

The input that neurons receive comes from other neurons. Neurons learn to recognize a pattern in that input and signal to following neurons how close the current input is to their learned pattern. Neuron learning occurs at the synaptic level. Active synapses that are part of an input pattern that causes the neuron to signal pattern recognition are strengthened. Patterns of input that cause the neuron to signal pattern recognition are learned.

Special support cells provide a myelin sheath that surrounds most of the axons in the human nervous system. The myelin sheath prevents signal degradation from the axonal membrane for the neuron signals propagating down the axon. Many myelin containing cells line up along a neuron's axon in order to coat the entire length of the axon.

Neuron cell bodies clump together and inhabit the same places in the body. These groups of neuron cell bodies are called nuclei in the Central Nervous System (CNS) and ganglia in the Peripheral Nervous System (PNS). The PNS is everything outside of the spine and brain case. The output axons from these groups of neurons also run together in the body. These multiple axonal paths are called tracts in the CNS and nerves in the PNS.

Each neuron functions very much like a voting booth counting both yes and no votes. Votes are received on the synapses of the dendritic input structure, the dendrites deliver the votes to the cell body where they are collected, the beginning of the axon counts the votes, and if there is a big enough majority in favor, the axon signals to following neurons. There are both yes type votes to count (excitatory) and no type votes to count (inhibitory). Individual inputting neurons are either yes voters or no voters and they never change.

Dendrites serve as the actual voting booth, synapses serve as the actual ballot and the ink involved is called a neurotransmitter. Dendritic shapes resemble tree limbs in their branching patterns and are commonly referred to as dendritic arbors. The number of dendritic inputs a neuron supports varies from one input in sensory neurons to up to 200,000 inputs in some learning type neurons.

Dendritic synapses are typically excitatory and are often, but not always, supported by dendritic spines. Synapses on the cell body are often inhibitory and are well positioned to prevent the neuron from firing. With billions of firing neurons in the brain, a large investment is made in inhibition in order to filter out noise.

The cell body is the polling location where the votes are collected. Each individual synaptic input vote arrives at the polling cell body. This takes a small amount of time and each vote is only valid for a certain amount of time. Each input vote starts out as a full vote and then decays away. If the yes votes out number the no votes by some amount, the threshold amount, the neuron signals this via its axon.

The neuron cell body also performs the entire normal array of cellular house keeping duties. Protein synthesis, cellular repair, energy production, etc. are normal cellular functions performed by the neuron cell body. Many of the neurotransmitters used as chemical messengers in axonal synaptic clefts are produced in the cell body and transported down the axon for use. One cellular function not typically performed by the neuron cell body is mitosis or cell division.

The portion of the output axon adjacent to the cell body is called the axonal hillock. This is where the votes are counted. A successful vote causes the neuron to fire and produce an action potential. The remaining axonal output portion of the neuron has only one job, transmit the output signal from the cell body over some distance and provide that signal as input votes to the input area of some other cells, neurons or muscle cells.

There are only two types of postsynaptic cells, muscles and neurons. In neuron to muscle signaling, each individual muscle cell receives input exclusively from one neuron. Neurons that output to muscle cells are called motor neurons. Motor neurons are the final neurons in output circuits that control your muscles and are contained mainly in the spinal cord. A motor neuron can output to a few muscle cells where fine control is required, such as the hand, or can drive many muscle cells in large muscles not requiring fine control. There is only one neurotransmitter involved, acetylcholine, and the neurotransmitter effect is always excitatory.

The axon signal, an action potential, is an all or nothing type of signal. A one-millisecond positive voltage spike propagates down the axon away from the cell body at a high rate of speed. The axon quickly recovers to its initial state and after a minimum resting period, is available to fire another pulse. If the electrical state of the axonal hillock is still above the threshold, the neuron will fire again. The neuron is a pattern recognition device and the strength of the pattern recognition is indicated by the frequency of axonal firings.

The length of a neuron's axon is dependent on the neurons location in the body and the body location of its output targets. In the brain these axonal connections tend to be very short. Some sensory neurons in your toe actually terminate in your brain stem.

In a large person that represents a distance of approximately six feet. An axon typically gives off many branches called axonal collaterals at various points along its path.

Myelin insulation greatly improves the speed of axonal transmission. The size of the axon also has a direct bearing on axonal signal speed, bigger being faster. A large myelinated axon is approximately 25 to 50 times faster in signal propagation than a small unmyelinated axon. Small unmyelinated axons are associated with neurons that propagate pain signals. These are perhaps the oldest sensory system neurons in an evolutionary sense.

Communication between neurons occurs at synapses and is chemical in nature. The chemicals involved are called neurotransmitters and there are many different kinds. Research has identified more than forty different types of neuroactive substances to date. This chemical signal is transformed into an electrical signal by an ion gating channel within the receiving synapse. All internal signaling within neurons is electrical in nature. The representation of votes internal to the neuron is electrical and the output signal that propagates down the axon is also electrical.

An ion is a molecule with a positive or negative charge. In the transformation from chemical synaptic signaling to inter-neuron electrical signaling, it is ions that provide the electrical charge. Yes type gating channels allow a particular type of positive charged ions to enter the cell and no type gating channels allow a particular type of negative charged ions to enter the cell. Each influx of charged ions changes the internal electrical state of the neuron. Individual influxes of charged ions, yes and no votes, typically contribute around .2 to .4 millivolts of positive or negative charge to the neuron. This electrical charge dissipates from the ion channel location, the synapse, throughout the cell interior and degrades with time.

Ion channel gates are selective for one ion type and these ions can have either a positive or negative electrical charge. The delay from receipt of the axonal pulse at the synapse to electrical effect in the receiving cell is about one millisecond.

After the axonal pulse, the neurotransmitter is destroyed or absorbed for reuse quickly to allow another signal to be transmitted.

The effect a particular neurotransmitter has on the receiving cell is dependent on the receiving ion channel gate type and not on the neurotransmitter itself. There are multiple receptor types for virtually all neurotransmitters. Signal effects on the postsynaptic cell can be either excitatory or inhibitory depending on the electrical charge of the ion admitted by the ion channel gate. The distribution of signaling in the human brain is about one half excitatory and one half inhibitory.

It is the sum total voltage effect of all of these inputs that will cause the neuron to fire if the inputs raise the cell body voltage by around ten millivolts. The signaling neuron's axon branches into many axonal collaterals and forms synapses with the dendrites of many other neurons. An action potential propagating down an axon causes each axonal collateral to release its neurotransmitter into its synaptic gap. The chemical neurotransmitter causes the receiving neuron synapse to physically open an ion gate channel into the neuron. An ion gate channel is made up of proteins that change their shape in the presence of the neurotransmitter to open a hole into the neuron. That open ion gate channel allows a particular type of charged particle into the neuron. The flood of charged particles allowed into the neuron's dendritic structure causes an electrical charge internal to the receiving neuron.

An action potential is electrical in nature and originates in the axonal hillock. A positive charge of ten millivolts across the axonal hillock's membrane causes membrane channels to open that allow positively charged particles to flood into the axonal hillock. That positive voltage change continually induces successively adjacent areas of the axon's membrane to open their channels and the incoming flood of positive charged particles propagates rapidly down the axon. After a millisecond these electrical input membrane channels close and electrical output membrane channels open that allow positive charge to flood out of the neuron. This positive charge inflow followed by outflow produces a large positive voltage spike that propagates rapidly down the axon.

In more detail, the axonal hillock fires by opening influx ion channels that allows positive sodium (Na+) ions to flood into the axon. The internal voltage of the axonal hillock is raised from a negative -65 millivolts to positive +35 millivolts very quickly. This large positive voltage change causes the adjacent area of the axon to open its sodium positive influx ion channels and so forth down the axon with each successive axonal area responding. This propagation of positive charge proceeds down the axon at a rapid rate, very much faster than the passive diffusion of charges in the dendritic region.

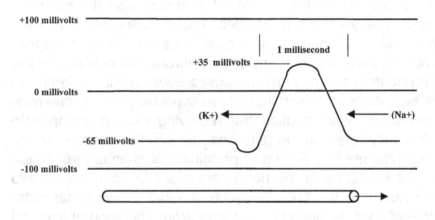

Figure 1-2: Axonal Action Potential

After around one millisecond, each axonal area shuts its positive influx ion channels and opens positive Potassium (K+) out-flux ion channels to allow the internal axonal voltage to return too normal. The axon voltage is driven slightly lower than the resting potential and the potassium out-flux channels close. The axon then stabilizes at the normal resting potential ready to propagate another pulse. The result of each axonal area turning on and then turning off is a one millisecond positive (+100mv) voltage pulse that propagates down the axon away from the cell body. (Ref. Fig. 1.2) This positive voltage spike arrives at each collateral ending of the axon and causes the

8

release of a neurotransmitter into the synaptic gap. This results in a corresponding yes or no vote being registered electrically in the following neurons.

How neurons and collections of neurons are altered to support memory and learning remains the main quest of neuron research after decades of effort. In 1949 Donald Hebb proposed the idea that synapses that were open or active when the neuron fired were somehow strengthened or enhanced. Synapses that opened or became active but did not cause the neuron to fire were somehow weakened or diminished. Synapses that never contributed to input patterns that caused the neuron to fire would weaken and disappear. He proposed this principle of synaptic modification as the basis of learned pattern recognition. This basic idea of synaptic modification enabling the pattern detection capability of the neuron is still a core principle of operative hypotheses today.

Lets repeat the core idea of that last paragraph. The modification of synaptic efficiency to improve the pattern recognition capability of neurons was proposed by Mr. Hebb in 1949 as the neural mechanism facilitating learning and his theory remains the operative hypothesis to this day. This modification of synapses supporting the learning of pattern recognition is an extremely important concept and forms a cornerstone of our analysis as to how your brain works.

We still do not completely understand how this works physically but how it works logically is pretty simple. Lets assume that you are a dendritic synapse supporting a yes type voting input. Every time the axonal collateral of your input neuron spits neurotransmitter at you, you open your ion channel and register a small electrical yes vote. If when you vote yes the neuron you are a part of fires, your next votes count more. Your vote is more important. It makes things happen. If when you vote yes the neuron does not fire, your next votes count less.

If your vote is part of a pattern of votes that causes your neuron to fire, your vote becomes stronger and your neuron is more likely to fire if that pattern of input votes happens again. That is how learning type neurons learn to recognize patterns.

Dendritic synapses that do not contribute to the firing of their neuron atrophy and disappear. Neural circuits that do not cause downstream pattern recognition atrophy and whole neurons die. Neural circuitry that contributes to pattern recognition grows and is strengthened.

Neurons in the learning areas of your brain are continuously participating in the rewiring of neural circuits. Non-useful neural circuitry is eliminated. Axonal collaterals are grown to increase input to neurons where the input does contribute to action potentials. This results in the actual rewiring of neural circuits and is the main mechanism in the brain's ability to rewire itself to recover from brain injury. The plastic nature of the brain and its ability to partially recover from trauma is remarkable.

Neuron cells have many similarities with muscle cells. You are born with an extremely large quantity of both muscle and neuron cells. There is a saying in weightlifting, "Use it or lose it". Muscle cells that are used actively grow in size and contraction power with that use. A casual glance at a serious bodybuilder is proof positive that we have more muscle cells than we normally use. An arm or leg that is immobilized in a cast for six weeks is the opposite proof of losing muscle mass with disuse. Muscle cells shrink dramatically in size and contraction capability with non-usage.

Use it or lose it applies equally as well to neurons. Neurons that support memory and learning, such as those in the cerebral cortex, respond to usage by improving their synaptic capabilities. Animals raised in enriched environments develop larger dendritic arbors, more neural mass and thicker cerebral cortexes than animals raised in impoverished environments. On the other hand, neurons that are not part of a functioning neural circuit die. Injury or disease in the brain has a domino effect. Neurons that received input from damaged areas atrophy and die. The neurons that received input from them then atrophy and die. An autopsy of a person with a serious brain injury can often show a massive loss of neuron cells in areas quite removed from the original injury. You produce vast quantities of neuron cells at birth and approximately one third of these cells will die by adulthood. Neuronal cell death is the norm for the human brain.

Of the body's muscle cells, heart muscle cells most resemble neurons. Heart muscle cells have electrical gap junctions between cells that are like the electrical gap junctions found in some neurons. These electrical gap junctions directly couple the electrical internal state of adjacent heart cells. The heart receives a neural input signal at one point and that electrical signal is propagated across the heart to every heart muscle cell via electrical gap junctions. Each heart muscle cell contracts as it receives this signal and this drives the entire heartbeat.

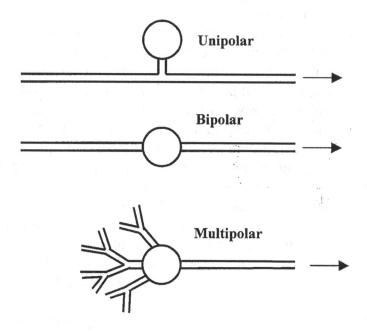

Figure 1-3: Types of Neurons

There are three types of neurons in the human nervous system and neuron shape is the distinguishing factor. (Ref. Fig. 1-3) The Unipolar neuron has both its input and output structures combined into one connection to the cell body. Both the input and output portions are axonal in their behavior and support action potentials. Unipolar type neurons are extensively employed in the nervous systems of invertebrates. In vertebrates, unipolar neurons are confined to sensory neurons whose cell bodies are contained in ganglia of the peripheral nervous system. When a sensory neuron detecting touch in the skin fires, an action potential in its sensory apparatus propagates to its axonal collaterals in the central nervous system.

Bipolar neurons support just one dendritic structure in addition to their axon. Bipolar neurons are rarely employed in the human nervous system. They appear as sensory input neurons in the retina and the olfactory sensory neurons of the nose. Multipolar neurons support two to many dendritic structures and are the predominant type of neuron of the human central nervous system.

The neuron is not a digital device. It is rarely completely statically on or off. The neuron is sensitive to particular patterns of input on its input synapses and signals that sensitivity by the frequency of its output pulses. Different kinds of neurons have different resting states with respect to their output patterns. Many cortical neurons and others are normally silent. Some neurons fire at a constant frequency and modulate that frequency up or down depending on their input. Other neurons involved in the control of rhythmic behavior such as breathing or walking fire regular bursts of pulses to affect their control.

The human brain is made up of many more cell types than just neurons. In addition to neurons and the normal organ support systems such as blood circulation, the brain contains glial cells and cells that contain myelin. These cells are unique to the nervous system. Glial cells, of which there are many types, provide nourishment and support for the neurons. The cells containing myelin wrap themselves around the axonal output-signaling portion of the neurons to provide insulation and improve signaling speed. Neither of these cell types is believed to provide any pattern recognition capabilities and we will not devote particular attention to either. Neurons represent around 10% of the cells and approximately 50% of the mass in the human brain and are the functional building blocks from which the brain is constructed.

Some neurons send chemical substances to large areas of the brain and nervous system. These substances are delivered directly into the blood or extra cellular matrix via varicousities along their axons. These neurons and the brain structures they build effect hormonal controls and alertness.

In addition to the typical ubiquitous chemical synapses, there are much less common electrical synapses called gap junctions. These gap junctions directly connect the electrical internal environment of one neuron with another as they do in heart muscle. Gap junctions seem to be employed in the human brain to synchronize the activity of groups of neurons. They exist in the neurons that make up the retina of the eye and also connect the dendritic arbors of neurons in a unique component of the thalamus.

Conclusion (Pattern Detection - What Neurons Do)

If the limitations of neurons define what a brain can be, lets examine how evolution has employed neurons in the architecture of the human nervous system. There are relatively few neurons that make up most neural circuits. Tactile sensory input to the brain involves just three neurons in series from the skin to the cerebral cortex. Motor output from the cerebral cortex has even one less neuron in the circuit. That is just five neurons in series in support of sensory input and motor output to and from the cerebral cortex.

Brain components that learn employ neural circuitry that is both regular and consistent. The neural circuitry that makes up these components employs a fixed number of neuron layers to accomplish the component's function. The regularity and consistency of neural circuitry within brain subsystems is remarkable. The overall design scheme of the human brain utilizes neurons with very large numbers of inputs and outputs, very few levels of neurons in any one neural path and enormous numbers of neurons in each level. This brain architecture is best described as a serially shallow, vastly parallel implementation. Apparently the best solution for the limitations of the neuron is do not depend on individual neurons but have lots of neurons vote at each level of every neural circuit.

Each individual input vote to the dendritic arbor of the neuron decays through membrane leakage of charged ions. Each full vote decays away within a short time window. In order for a neuron to fire, a number of excitatory inputs must be received by the dendritic arbor within a short window of time. This decay of each synaptic input causes the neuron to be most sensitive to patterns of input that are synchronous in time. A strong synchronous pattern of inputs will always cause the neuron to fire. Input patterns that are spread out over time have less probability of causing the neuron to fire. This propensity of the neuron to be more sensitive to synchronous input is a very important part of our analysis as to how your brain works.

One of the striking things about learning type neurons is their large number of input and output connections with other neurons. A pyramidal neuron cell in the human cerebral cortex has typically 60,000 inputs and 20,000 outputs. That is an average of 80,000 interconnections for just one neuron. The wiring complexity of vast numbers of microscopic neurons with an average of 80,000 interconnections each cannot be overstated. Couple this wiring complexity with chemical inter-neuron communication employing over 40 different kinds of neurotransmitters, multiple types of ion input gating per neurotransmitter and different electrical effectiveness of synapses depending on their neural location and the true complexity of the human brain begins to come into focus.

So this is the neuron we have to work with, our brick. Now it is time to find out what neurons build.

Chapter 2

Your Brain Building
The Basic Floor Plans

Introduction

Neurons combine to build neural circuits, neural circuits combine to build neural components, and neural components combine to build brains. Chapter one hopefully articulated how neurons build neural circuits. In this chapter you will gain an introductory understanding of the various components that neural circuits build. We will discuss each component physically, what neural circuits it contains, how it interconnects with other neural components and what function it provides.

We are going to take a bottom up approach in our analysis. In an attempt to comprehend a complex system, it is customary to break the system down into its respective parts in order to understand the function of each with respect to the whole. This is the main theme of this chapter.

The human nervous system is partitioned both physically and logically. The physical partition divides the whole into the Central Nervous System (CNS) and the Peripheral Nervous System (PNS). The central nervous system is everything contained within the spine and skull, the brain and spinal cord, and the peripheral nervous system is everything outside of the spine and skull, the vast network of neurons providing sensory input and controlling your body.

The logical partition divides the whole into the somatic nervous system and the autonomic nervous system. The somatic nervous system controls voluntary movement and drives the conscious portion of you. The autonomic nervous system is your unconscious control system driving everything else.

Everything else turns out to be quite a lot. Control of your digestive tract, your blood pressure, your body temperature, kidney function, liver function, breathing, heart rate, sweat glands, and much more. The list goes on and on. The unconscious autonomic nervous system even carries much of the load for proper operation of skeletal muscles. Control of muscle tone, antigravity function, posture, and balance are autonomic nervous system responsibilities. The autonomic nervous system is completely made up of hardwired non-learning neurons and neural circuits.

These somatic and autonomic control systems are highly interdependent and interactive. The autonomic nervous system is the more primitive of the two and the neural structures in the brain stem that drive this system are referred to as your reptilian brain. These structures exist in reptiles for the same purposes and provide exactly the same functions as they provide for you.

A Tour of Your Nervous System

In order to provide a more familiar context for our discussions of neural components, we are going to tour the CNS as if it were a five-story building. (Ref. Fig. 2-1) Think of it as a brick building. The first floor, the spinal cord, looks like an elevator, represents the bottom vertical two thirds of the CNS and contains 31 vertebrate that support 31 pairs of spinal nerves. These nerves connect the CNS with the body via input and output axons. The spinal cord houses three distinct groups of neurons performing three distinct functions. Motor neurons output to every skeletal muscle cell in your body and drive muscle contraction. Sensory input path neurons receive input from sensory neurons in your body and pass that input up the spinal cord to your brain. Large numbers of interneurons exist within the spinal cord for modulation and interconnection of sensory input and motor output.

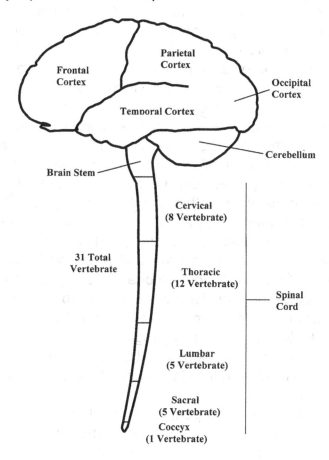

Figure 2-1: Central Nervous System (CNS)

The second floor contains the brain stem located on top of the spinal cord. The brain stem exerts control over the unconscious autonomic nervous system and is made up of three distinct rooms. The medulla oblongata sits on top of and closely resembles the spinal cord in structure. The medulla oblongata regulates autonomic functions such as blood pressure. The pons sits on top of the medulla oblongata and also regulates autonomic functions as well as providing the interconnection paths between the brain stem and the cerebellum. The midbrain is the top structure of the brain stem and plays a large role in the control of eye movements.

The second floor also contains a fourth room, the cerebellum, located posteriorly adjacent to the pons. The cerebellum is essential in the learning of coordinated movements. Your entire cerebral cortex serves as input for the cerebellum's function. Cerebellar output targets the portion of your cerebral cortex that directly drives the motor neurons controlling your muscle cells.

The third floor of your brain building is located directly above the brain stem midbrain and contains two rooms, the thalamus and the hypothalamus. The thalamus acts as the central switchboard for all sensory information destined for the cerebral cortex. The senses of touch, sound and vision each have connections within the thalamus specific to that type of input. Thalamic output terminates in different areas of the cerebral cortex specific to each type of sensory input. The hypothalamus sits just below the thalamus and is the main output center controlling the hormonal state of your body.

The basal ganglia consist of a group of six rooms contained on the fourth floor. These rooms lie above, to the side of, and below the thalamus. The basal ganglia receive essentially the same massive input from the cerebral cortex that the cerebellum receives. The basal ganglia play a central role in the control of neural output.

The top fifth floor, the penthouse, contains your cerebral hemispheres. This is the walnut like exterior surface of your brain that we are all familiar with. The cerebral cortex is a thin sheet of highly structured neurons that covers the entire human brain. The cerebral cortex is the latest evolutionary addition to the brain and is in every sense the brain structure that makes you human. The oldest portion of the cerebral cortex is called the hippocampus. It is impossible to store memories without a functioning hippocampus.

The newer areas of the cerebral cortex are divided into four different lobes with different and distinct responsibilities. The frontal cortex receives emotional input, stores the patterns that define your behavior and drives the motor neurons that control skeletal muscles. The parietal cortex receives sensory input from the skin of your body and contains the patterns that define your neural virtual world. The occipital cortex receives visual input from your eyes and creates the pattern inputs that define your visual world to all of the other cortical lobes. The temporal cortex receives auditory input and contains the patterns that define all of the objects you recognize. The temporal lobe is especially critical in the implementation of language.

The Peripheral Nervous System (PNS) contains all of the nerves that exit the CNS, all of the ganglia collections of neurons in the body and all of the body's specialized sensory neurons. Spinal nerves exit each segment of the spinal cord and contain both input axons from various sensory neurons and the output axons of motor neurons that terminate on the skeletal muscle fibers of the body. There are an additional twelve pairs of cranial nerves that exit the CNS in the head region. The cranial nerves are the main input and output nerves of the brain stem that controls the autonomic nervous system.

This tour is designed to trace a complete round trip from sensory input to neural output. Our tour begins with a look at the various neural input systems that define your external world. After our discussion of input, we will visit the first, second, and third floors. We will skip the fourth floor on the way up because that's what sensory neural input does to visit the fifth floor cerebral cortex. Our trip down will stop on the basal ganglia fourth floor. We will end our tour with a discussion of neural output. Neural output in humans encompasses all of human behavior. We will confine our discussion in this introductory phase to the control of skeletal muscles.

Neural Input – Building Your Virtual World

External world input comes in six flavors. In each of these input types you have specialized neurons that convert the representation of the external world into action potentials that can be interpreted by your nervous system. Sensory input receptors are specialized neurons that are essentially energy transducers. They transform one kind of energy into an action potential in the axon of the receptor neuron. The sense of touch, pain, temperature and body position are referred to as somatosensory input. Light and sound energies are interpreted by your visual and auditory systems. The sense of balance involves a combination of gravity and motion detection. Gustatory taste sensation and olfactory or smell involve chemical detection and discrimination. These six types of input are transmitted into the CNS via nerves.

Sensory input paths are precisely wired. The actual two-dimensional alignment pattern of sensory inputs is exactly reflected in the physical alignment of the axons that make up the input path. This has the effect of maintaining an accurate physical representation of the sensory input throughout the input path. Somatosensory input from the skin of your body is maintained perfectly throughout the entire input path to terminate as a precise two-dimensional representation of your skin in the parietal cortex. The input path for vision is a two-dimensional array of the images that fall on the retinas of your eyes. The audio input path is likewise a precise array based on frequency.

The amount of sensory input continuously received by the CNS is extremely large, you cannot possibly pay attention to all of it. Inhibition in order to select the most critical input is extensive. Inhibition starts with the second neuron in the sensory input path and exists at all other neurons in the path. A great deal of sensory input never reaches your cerebral cortex and serves as input to your unconscious autonomic nervous system functions. Fully two thirds of the somatosensory input from your body is diverted to the cerebellum and brain stem.

All input paths synapse on neurons in the third floor thalamus, which acts as the gatekeeper for input to the cerebral cortex. Inhibition is especially intense in the thalamus. Sensory input paths, with minor exceptions, cross over to the opposite side of the body before reaching the thalamus. Sensory input from the left side of your body is received in the right side of your cerebral cortex and input from the right side is received in the left side of your cerebral cortex. All sensory input types have a discrete area of primary sensory input cortex where they are first received.

How do you feel? You feel because specialized neurons in your skin and body fire to signal the detection of energy. Neurons in your skin detect energy exterior to your body and neurons inside your body detect energy caused by the action of skeletal muscles. Secondary neurons in your first floor spinal cord or second floor brain stem pattern detect this input and project it on to third level neurons in the third floor thalamus. The thalamic neurons project to the parietal primary sensory cortex (S1). It is the fifth floor cerebral cortex that allows you to be conscious of this energy, allows you to feel.

Somatosensory input is concerned with the state of the body and the body's contact interface with the external physical world. The receptor neurons that detect these input energies are unipolar neurons with a single axonal type appendage. The external ends of this axon appendage are specialized energy transducers designed to detect a certain type of energy. The receptor neuron body is contained within a group of neurons called a spinal ganglion on the side of the spinal cord. The internal segment of the receptor neuron axonal appendage supports multiple collateral terminations in the spinal cord or brain stem as the first synapses of the input circuit.

There are three basic types of neuron receptors within the somatosensory input system. Exteroceptors detect energy from the external environment via the skin and subcutaneous layer. The simplest receptor of this type of is the free receptor that detects pain and temperature. Perifollicular receptors are associated with hair follicles of the skin and detect tactile energy associated with movement of their associated hair. There are many other more sophisticated exteroceptors called encapsulated receptors that detect vibration, pressure, two-point discrimination, and all of the other subtle aspects of your sense of touch.

The second class of receptors is proprioceptors that provide information about body position and the contraction status of muscles. Your entire body contains specialized Golgi tendon organs and muscle spindles that signal either position or change of position to the CNS. These neural input devices accompany every tendon and skeletal muscle in your body. This is a large amount of neural input defining the mechanical state and movement of your body. In addition to the cerebral cortex, this input is mainly destined for the second floor cerebellum where it is used to guide and control learned movements.

The final class of input receptors is called interoceptors and exists in the walls of organs and blood vessels. These specialized receptors are a vital part of the servo control loops that make up the autonomic nervous system.

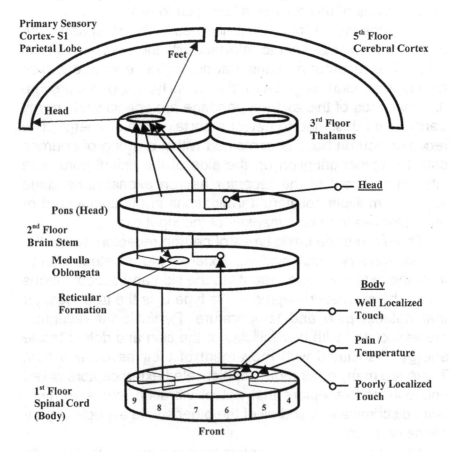

Figure 2-2: Somatosensory Input Paths to the Cerebral Cortex

Somatosensory input paths involve three neurons between receptor signal and cerebral cortex. (Ref. Fig. 2-2) The receptor neuron synapses with multiple secondary neurons in the first floor spinal cord or second floor brain stem. Those secondary neurons synapse with multiple third level neurons within the third floor thalamus. The final synapse in the input paths is in the primary sensory cortex (S1) in the parietal lobe of the cerebral cortex.

Inhibition of somatosensory input begins with the second neurons. The receptor first neuron, located in the PNS external ganglion, transmits its signal faithfully without inhibition. Secondary neurons in the spinal cord or brain stem receive inhibiting input directly from the primary sensory cortex (S1) and from the collaterals of other input fibers in a contest for attention. The neurons in the nuclei of the thalamus that pattern detect somatosensory input again receive inhibiting input. All of this inhibition at every level is designed to filter the vast incoming information flow to only the most important.

Only the most intense input signals or the signals the brain wishes to concentrate on are not inhibited. If you wish to concentrate on your left big toe, you can become conscious of the somatosensory input from that area. Inhibition of input from your big toe is suddenly removed and inhibition increased for other somatosensory input. The S1 cortex facilitates that wish.

Somatosensory signals are physically positioned in a precise two-dimensional linear map of your body in all nerves, ganglion, neural tracts, nuclei and cerebral cortex. Different types of somatosensory input utilize separate and distinct paths from receptor neuron to cerebral cortex.

Pain and temperature are the first body sensations to have evolved and remain the least advanced. The receptors involved are called free receptors and are just the unmyelinated ends of unipolar neuron endpoints near the external surface of the body. Damage to the body or head causes damage to these free nerve endings and causes them to fire repeatedly.

Secondary neuron axons carrying pain and temperature information to the thalamus are unmyelinated and are therefore very slow in transmittal speed. The ability of humans to finely localize pain is the most advanced of all vertebrates but remains limited. Collateral connections of the pain input fibers within the reticular formation of the second floor brain stem alerts the rest of the brain quickly to pain signals. Of all somatosensory input signal types, pain has evolved the most advanced inhibitory processes.

Your keen sense of localized touch is facilitated by a variety of highly evolved receptor type neurons specialized for vibration, pressure, touch and two-point discrimination. These receptors are located in different layers of the epidermis and subcutaneous layer and are arrayed in different densities depending on the need for fine resolution of touch. The fingers and tongue have the highest density of these receptors and therefore the greatest sensitivity to touch.

Somatosensory thalamic neurons are arranged in small cylinders, each dedicated to a specific type of input from a distinct area of the body or head. The outputs from these cylinders are directed at columns of neurons in the primary sensory cortex that are likewise dedicated to the same input type from the same localized region of the body.

The primary sensory cortex (S1) is a strip of cortex located in the front of the parietal lobe that runs across the top of your head from ear to ear. S1 has a precise representation of all somatosomatic body inputs in a body map. The left side of the body is represented on the map of the right primary sensory cortex and vise versa. In all vertebrates the map of the body is similar. The feet are located in the medial, top of S1 and the head is located in the lateral lower section. The body is always facing backward with the area of cortex dedicated to the various body parts in direct proportion to the amount of input from that body part.

The retina at the back of the eyeball contains sensory neurons that transform light energy into action potentials destined for the primary visual cortex (V1) in the occipital cortical lobe. The pattern of input produced by each eye represents a visual snapshot of your world. This snapshot is a two-dimensional, static image. Think of each eye as a one-mega pixel digital camera that takes a continuous series of static pictures and sends them through the thalamus to V1. What we perceive visually is a three dimensional dynamic world. The conversion from a two-dimensional static world to a virtual three-dimensional world with moving objects takes place in your brain.

The retina breaks the image you see into separate dot-like representations. Each of these dot patterns is made up of two concentric circles. On center dots consist of a light inner circle inside a dark outer circle and off center dots consist of a dark inner circle inside a light outer circle. Your retinas provide separate and complete images to your brain made from on center dot patterns and off center dot patterns. These dot matrix representations of what you see are sent unaltered to the fifth floor primary visual cortex (V1).

The neurons in (V1) pattern detect this duel image input of concentric circle patterns to detect lines and edges in every area of the visual field. After these two transformations, the entire visual image consists of what lines and edges of various orientations exist in each small area of the visual image. From this line and edge representation, your brain reconstructs your virtual world.

The image that falls on the retina of your eye is first transformed into concentric circles and then transformed into a collection of all of the lines and edges that are detected in the image. This seems like a lot of work when we had a perfectly good visual image to begin with. Why bother? All of this pattern detection to transform the visual image into a line and edge representation is what allows you to see objects. You are very good at seeing objects. Your entire visual image is made up of things. The visual image of an object causes its neural object pattern to light up in your cerebral cortex as you recognize objects and put them together to build your virtual world. Visually detecting and identifying objects is what your visual input system has evolved to accomplish.

Your visual system allows you to see your world much better than any other living thing on earth. Humans are visual animals and your visual abilities are astounding. The visual input path and visual primary cortex are the most studied portions of the human brain. We know more about how you process visual information than any other input type. Lets have a look at what we know.

The center of the retina is called the fovea. The fovea has the highest density of visual receptors and is the location of highest visual acuity. Humans continually move their eyes in movements called saccades to bring the point in the visual field they wish to examine onto the fovea.

The visual input path is virtually identical in all mammals. Human vision, while identical in path and functional characteristics with all other mammals, has evolved to a unique level of capability. In the human, the optic nerve from each eye contains approximately one million axons. The audio nerve contains approximately 30,000 axons. Your visual input tract contains more input axons than all of the somatosensory input entering the spinal cord concerning your entire body.

Primates are the only mammals that perceive colors. All other mammals see the world in gray scale. The ratio of primary visual cortex neurons to visual path neurons in the thalamus indicates the level of pattern detection applied to visual information in the primary visual cortex. In the rabbit this ratio is twenty to one (20:1). In the monkey it is one hundred and forty five to one (145:1). In the human primate it is nine hundred to one (900:1).

The human visual input system is in most ways similar to all of the brain's input systems. A two dimensional view of the external environment is focused by the lens of the eye on the retina. Specialized input receptors in the retina, rods and cones, detect and signal receipt of the light. This two dimensional representation of the three dimensional external world is linearly maintained by the visual input path through the thalamus to the cerebral cortex. A unique characteristic of the visual input path is the lack of neural inhibition. The cerebral cortex receives the entire two-dimensional circle dot pattern representations of the image falling on the retina.

Your visual system superiority is matched by your ability to remember what you have seen. People shown a series of pictures for five seconds each, up to 10,000 pictures, and then later asked to identify from two pictures the one seen before, score in the high 90th percentile. Your storage capacity for visual memories seems unlimited.

Neurons perform pattern recognition and the neurons performing pattern recognition of visual information are extremely accurate with only a single brief input exposure. Everyone has experienced visiting a place once visited years before and recognizing it immediately or remembering what it looks like. Considering the amount of visual information you input in a day, in your life, your visual pattern recognition abilities are amazing. So, now that we know that you see very well and remember what you have seen very well indeed, lets examine how you see.

The eye is a light gathering device designed to clearly focus an image at some selected distance from the eye onto the retina at the back of the eye. All mechanical aspects of eye function are under neural control derived mainly from the brain stem. Focus via lens shape, amount of light allowed via pupillary constriction, saccade to movement, etc. are all controlled by reflexive circuits in the brain stem and are similar in all mammals.

There are three neuron cell layers that make up the retina. (Ref. Fig. 2-3) The first layer located at the extreme back of the eye is made up of the actual receptors, rods and cones. The second layer contains interconnecting neurons and the third layer consists of ganglion neuron cells. The retina is actually constructed backwards. The rod and cone receptors are at the back of the eye covered by the bipolar and ganglion layers. Light must pass through the ganglion and bipolar layers to reach the rods and cones. The axons of the ganglion cells that combine to form the optic nerve are unmyelinated and thus transparent to light. In the fovea region the ganglion cell bodies and axons are shifted out of the way to minimize distortion.

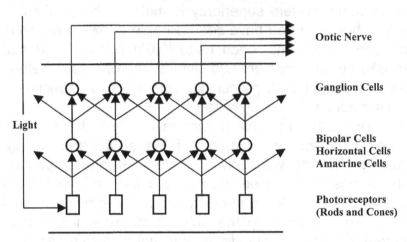

Figure 2-3: The Retina

There are approximately 110 million light sensitive cells in the retina of each eye. Most of these are rods, approximately 100 million. Rods are specialized to detect low intensity illumination important for night vision. Rod cells supply input to many ganglion cells and one ganglion cell receives input from many rod cells. The other light receptor cell type is the cone cell. There are approximately 7 million cone cells in the retina. Cone cells provide color vision and detection of fine detail and are more concentrated in the fovea.

The ganglion cell is the first point in the visual input path where a pattern is detected and that pattern is a circle with a contrasting center and surround. The surround portion of the pattern is more important than the center or dot portion and the lack of a proper surround will inhibit the cell's firing regardless of the dot.

Ganglion cells in the retina are never silent but fire continuously at some frequency. That frequency is modulated up or down depending on how close the ganglion cell's input matches the circular pattern it is specialized to detect. Ganglion cells also increase their firing when their detected circular pattern is removed from their receptive field. Ganglion cells are very complex neurons that signal how close their input matches their optimum circular pattern and also signal changes away from that optimum pattern in support of movement detection. The cells of the retina also contain electrical gap junctions that connect the internal electrical state of these cells, similar to heart muscle. This causes the firing of the retinal output cells to fire in unison and project a synchronous two-dimensional pattern of the visual image to the primary visual cortex.

There are two types of ganglion cells, M and P cells, and both of these types have on and off center type cells. M type ganglion cells have large receptive fields and are involved in the detection of motion. P type ganglion cells have smaller receptive fields and are involved in color discrimination and detection of fine detail. The input paths of these two cell types are separate and distinct through the thalamus to the primary visual cortex. There are also some specialized ganglion cells that detect overall illumination to enable the pupillary reflex controlled by the brain stem.

The optic nerve is made up of approximately one million ganglion axons from ganglion neurons pattern detecting 110 million receptor neurons. There is approximately a 100 to 1 compression of receptor cells to nerve fibers in the retina. Most of these fibers provide input to the primary visual cortex via the thalamus and to the midbrain portion of the brain stem. The midbrain implements eye saccades by controlling head, neck, and eye movements to bring the eyes to focus on objects of interest. In the human, the midbrain works in conjunction with the frontal cerebral cortex to allow conscious control of eye saccades. Some visual input fibers connect with nuclei in the second floor brain stem to enable visual reflexes and some connect with the third floor hypothalamus to support your circadian rhythm.

The image of the left visual field in both eyes is combined and provided to the right primary visual cortex and the same function is provided for the right visual field. In this way the image from the right is provided to the left-brain and vice versa. (Ref. Fig. 2-4)

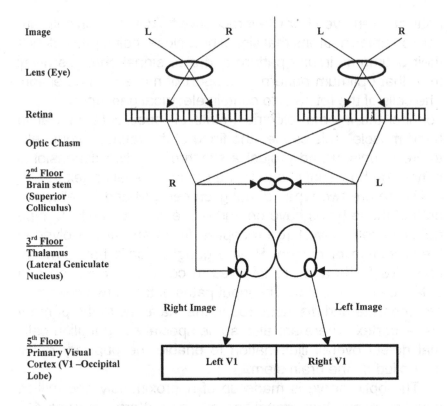

Figure 2-4: Visual Input Path

The neurons of the primary visual cortex (V1) in the occipital lobe at the extreme back of the head receive the output of the thalamus. The receiving neurons of V1 are interconnected to perform pattern detection of lines and edges and not circles. The primary visual cortex detects lines and edges of every possible orientation in each area of the visual field built from patterns of circles within that area of the field.

The human visual system is concerned with the contrast between the various retinal fields rather than with the light intensity. The detection of objects constructed from lines and edges depends on the detection of variation of intensity rather than the intensity itself. The line and edge patterns detected in V1 are passed on to other areas of the visual cortex that pattern detect ever more complex groupings of this input to finally enable objects within the visual field to be recognized.

The collection of objects recognized visually is a major part of your virtual world contained within your brain.

We humans don't hear nearly as well as we see. Compared with other mammals, human auditory capabilities are not great. An interesting feature of your audio input system is the amount of neural circuitry dedicated to pin pointing the source of sound. However that circuitry and capability are identical in all mammals and are not unique to humans

The neural input system related to sound has evolved to perform localization of sound. Determining from where a threatening sound is emanating has obvious survival implications. Localization of sound direction is determined by the difference in timing and the difference in sound intensity received between the two ears. Considering that your ears are extremely close together, the difference in time that a sound signal is received and the degradation in volume between the two ears is very small. Unique neural circuits in the brain stem are finely tuned to allow precise sensitivity to the location of sound generation. That neural circuitry has the ability to locate sound within one degree of a 360-degree circle.

Of all the input senses, hearing involves the greatest amount of mechanical equipment utilized to receive the input energy and convert it to action potentials. The ear captures and directs sound vibrations into the ear toward the eardrum. The eardrum covers the outer opening between the middle ear cavity and the outside world. Sound energy causes the eardrum to vibrate sympathetically with the energy spectrum received and this energy is transferred across the middle ear by three small bones to the oval window of the inner ear. The oval window covers the opening to the cochlea. This mechanical system of two drum type membranes interconnected by three moving bones has been exquisitely tuned by evolution to support the great range of frequencies and amplitudes detectable by the auditory input system.

The cochlea is a bony structure often compared to a snail shell or the horn of plenty. It is made up of two fluid filled tubes that make two and one half turns. The cochlea contains the organ of Corti that runs down its middle. The organ of Corti is where sound energy is converted into action potentials. The organ of Corti supports hairs that are innervated by the input fibers of spiral ganglion cells similar to the innervation of the hairs on your body. The cochlea is fluid filled and vibrations of the oval window are transferred via movement within the fluid to the hairs along the organ of Corti. The movement of these hairs causes impulses to fire in spiral ganglion neurons innervating the hairs. These action potentials are transmitted to the brain stem and on to the cerebral cortex.

Auditory input is received in the second floor brain stem by many brain stem nuclei and projected to the third floor thalamus. The thalamus projects to the primary auditory cortex located in the top of the temporal cortex on the side of the head. You can hear frequencies from 20 to 20,000 cycles per second, a thousand-fold range. You have the ability to discern sound vibrations a million times stronger than the faintest sound you can detect.

The auditory neural input system is similar to other input types in almost all respects. The main difference is that many neural structures receive input from both ears and there is extensive input crossover at all levels of the auditory system. The auditory system is linearly organized according to the frequency of sound waves.

Frequency linearity begins with the snail shaped cochlea and the organ of Corti. Each position along the organ of Corti is tuned to a different frequency. From the wider, thicker base to the narrower, thinner apex, low to high frequency audio energy is selectively detected. The inner hairs vary in length and thickness in such a way as to mechanically resonate with particular frequencies and the neuron cell is tuned electrically to be sensitive to the particular frequency for that location. This coupled mechanical and electrical sensitivity to a particular frequency acts like a sensitive tuned amplifier for that frequency. This frequency sensitivity varies linearly along the length of the organ of corti to support your entire auditory frequency input capability.

There are approximately 3000 inner hairs in the organ of corti innervated by approximately 30,000 spiral ganglion neurons. Each spiral ganglion neuron innervates only one hair for an average of ten input neurons per hair. Each individual neural input from the organ of corti is signaling information pertaining to one particular frequency. This array of 30,000 auditory inputs is arranged tonotopically from the organ of corti to the brain stem. Tonotopical linearity is maintained in the brain stem and thalamus and there are tonotopically arranged maps in the primary auditory cortex.

It's really hard to say if anything tastes good to my dog. She eats anything and everything I put in her dish. Since there are no neural design differences between her gustatory system and mine, I assume she tastes everything pretty much as I do. I do know that her sense of smell is vastly superior to mine. Her neural olfactory equipment is in every sense more advanced than mine, including the amount of her cerebral cortex dedicated to olfactory input. Humans are not superior to other animals in chemical perceptual abilities and are inferior to many vertebrates when it comes to sampling the chemical makeup of the air we breathe.

Your chemical neural input capabilities include two separate systems, gustatory or taste and olfactory or smell. These chemical systems appear to be the oldest, most primitive input sensory systems with olfactory being the very oldest. These chemical systems signal the detection and strength of chemicals in the environment and are not organized linearly with respect to input. These old input systems are closely interrelated with your third floor hypothalamus that is involved with emotion, food and sex.

The gustatory input path begins with taste receptors located in structures called taste buds. These structures resemble a flower in shape and are concentrated in the tongue. These taste buds contain input cells specialized in the detection of chemicals in food that produce the sensations of sour, sweet, bitter, and salty. These different specialized receptor cells are concentrated in various different parts of the tongue and mouth. These receptor cells have no axons and are innervated by receptor neurons. There are around 40 to 50 receptor neurons per taste bud with each receptor neuron innervating several receptor cells. This many to many design supports detection of overall concentration of particular chemicals and not where in the mouth they occur. Receptor neuron outputs project to the thalamus and on to the cerebral cortex. Your chemical taste input path is physically closely associated with the somatosensory input path from the tongue and mouth.

Olfactory input begins with chemical neural receptors in the lining of the nose. The axons of these chemical receptors cross the midline and project to the olfactory bulbs located just below the frontal cortex. These receptor neurons are unusual in that they die and are replaced approximately every 60 days. Division of stem cells forms new receptor neurons and the new neurons grow their axons to the olfactory bulb. The olfactory bulb must continually form new dendritic connections with new receptor neurons. The olfactory bulbs project directly to old cortical structures without first synapsing with the thalamus, the only input system to bypass the thalamus in this way. Subsequent connections of the olfactory input path do involve the thalamus that then projects olfactory input to the frontal cortex where your conscious sensations of smell occur.

There is one final neural input system that facilitates your ability to balance your body. This input comes from sensory organs in the inner ear that detect head movement and static position of the head relative to gravity. These organs contain hairs similar to the organ of corti for conversion of motion to action potentials. Neural input concerning head movement and position terminates in the brain stem where it is used to help control posture and movement. Very little of this input reaches the thalamus and cerebral cortex and your conscious awareness of balance is limited.

Floor #1 – The Spinal Cord (Your Brain's Input / Output Channel)

As the name suggests, all vertebrates have a backbone and all backbones contain a spinal cord. In very primitive vertebrates, the spinal cord is essentially most of the brain they have. What is interesting about these simple organisms is not their lack of capabilities but that the capabilities their spinal cord does provide are also provided by the spinal cords of humans. The first floor of your brain building is very old and has not been redesigned, redecorated, or even repainted, ever. What has been added is the ability to allow the floors above to take control of output and inhibit input.

The spinal cord is the oldest and most primitive portion of the central nervous system. The shape and size of the spinal cord varies over its length but it is generally a circular shape with a four-leaf clover of neuron gray matter surrounded by axonal white matter. The spinal cord supports all input from the body below the head and contains all of the motor neurons that innervate the skeletal muscles of your body.

A cross section of the spinal cord reveals two distinct areas, gray matter made up of neuron cell bodies and the white matter of myelinated axons passing vertically through that section of the cord. We will delineate spinal cord functionality by dividing the spinal cord into twelve pie shaped sections with sections one and twelve at the back or dorsal side. (Ref. Fig. 2-5) Sections two and eleven are called dorsal horns and consist of neurons receiving sensory input. Sections five and eight are made up of motor neurons driving muscular output and are called the ventral horns. The dorsal, ventral, and lateral sections are made up of white matter. Gray matter neurons concerned with the visceral nervous system are contained in lateral horns at the side of the spinal cord in the thoracic region and just below it. The size and shapes of these various sections vary over the length of the spinal cord.

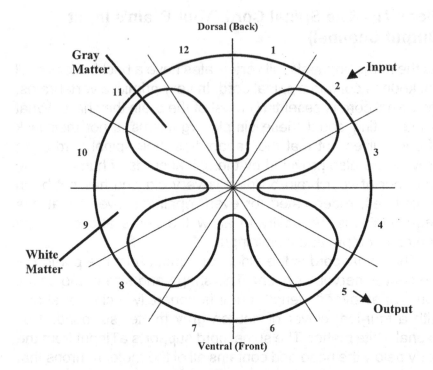

Figure 2-5: The Spinal Cord

The spinal cord supports input and output interfaces and provides rudimentary interactivity and reflexes between the two. Interfaces with higher level neural structures allow the inhibition of sensory input. This supports the selectivity necessary to filter input to a manageable level. Every level of the human nervous system above the spinal cord contributes neural output that affects skeletal muscles. Interneurons in the spinal cord integrate all of this higher-level skeletal control and provide it to motor neurons.

Your neural input circuits have evolved over time. Very old types of input terminate on neurons in the spinal cord before ascending to the brain, reflecting a period when the spinal cord was the only brain available. Sensations of pain and temperature seem to be the oldest detected sensations and these primitive inputs terminate on secondary neurons in the spinal cord. The axons of these secondary neurons cross the midline and ascend to the brain. Input concerning muscle positioning and tendon tension is received by secondary neurons in the spinal cord and projected to the cerebellum.

Newer types of input bypass the spinal cord to interface directly to the brain stem, reflecting a later period of evolution. Recently evolved inputs of finely localized touch enter the spinal cord and ascend without crossing the midline to the brain stem with limited collateral connections to spinal cord interneurons.

Spinal reflexes, inherited from very primitive organisms, are completely operational in humans. A sharp pain in the hand or foot causes interneurons to signal motor neurons and extract that extremity, just like a coral retreating into its shell. These types of reflex motor actions are exhibited in all sections of the spinal cord.

When a group of motor neurons is induced to fire and contract a muscle group, interneurons automatically inhibit the opposing muscle group's motor neurons from firing. This opposing muscle system is extremely important for coordinated movement and is exclusively a property of the spinal cord. Interneurons also support input from the brain stem facilitating motor neuron control for higher reflex actions. A great deal of the motor control for rhythmic actions such as walking are provided by the interneurons of the spinal cord.

The spinal cord is the oldest form of primitive brain. All of the functions of that primitive brain are still operational in the human spinal cord. Your first floor spinal cord provides a first level of input and output neural interconnection between the CNS and the body. It also provides a complex set of functions that enable the control of voluntary movement by higher brain functions. Many animals exist with nervous systems that are not much more than a spinal cord. Your spinal cord performs functions similar to theirs with additional interfaces provided for higher-level control.

Floor #2 – The Brain Stem (Your Reptilian Brain)

The second floor of your brain building houses four rooms and a common hallway. The first three rooms are stacked on top of each other with the common vertical hallway running down the middle of the rooms. A fourth large room is in back of and adjacent to the second room. Each of the first three rooms represents an evolutionary extension as organisms added neural material and corresponding capabilities to the head end of the spinal cord (Ref. Fig. 2-6). The common hallway is called the reticular formation. The fourth room, the cerebellum, is the first structure of our tour to have the capacity to learn. The oldest part of the cerebellum allows the learning of balance. Animals first needed to learn balance when they began to move about in their environment. The cerebellum and neurons that support learning have existed for a very long time.

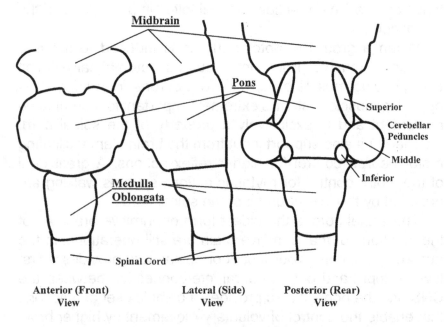

Figure 2-6: The Brain Stem

The brain stem is referred to as your reptilian brain. As this description implies, the brain stem comprises the major portion of the brain of a reptile. If you want to know what a brain stem is capable of, watch a reptile. Actually, a reptile is capable of quite a lot, eating, breathing, digestion, detecting food, sex, and a wealth of other behaviors that have allowed reptiles to survive for hundreds of millions of years. Snakes and lizards display really quite complicated behaviors and their brains, mostly the

38

equivalent of your brain stem, direct that repertoire of activity. The brain stem has responsibility for everything that it takes to keep you alive and a functioning brain stem is absolutely required to maintain life.

Approximately 10% of the neural material that makes up your central nervous system is contained within your brain stem. The brain structures above the brain stem that make all of human intelligence possible are for naught if the brain stem is not keeping your body alive and functioning. The work of the brain stem is unconscious and automatic with many of the brain stem's duties implemented as normal feed back systems. The brain stem also contains nuclei that produce quantities of certain substances that are delivered throughout the brain and spinal cord. In this manner the brain stem exerts control over the entire brain in determining your level of alertness. A damaged human brain stem usually results in coma and eventual death.

The brain stem is physically an extension of the spinal cord that extends into the head. A cross section of the brain stem closely resembles the spinal cord and its development is also very similar in the embryo. All of the functions that the spinal cord performs for the body are performed for the head by the brain stem. Head related somatosensory and proprioceptive input and control of motor output are exactly analogous. The senses of vision, hearing, taste, smell, and balance cause the brain stem to have the necessary neural capabilities related to those inputs as well.

The medulla oblongata is the first primitive brain to evolve on top of the spinal cord. Its main function is the control of the viscera or organs of the body. Neural input from the body's organs terminates here and output control destined for the smooth muscles that make up those organs originates here. The neurons and neural circuits that make up the medulla are all hardwired.

Body input bound for the third floor thalamus and all neural output bound for the skeletal muscles of the body crosses the midline between the left and right sides in the medulla. The reticular formation, the brain stem's common hallway, makes its first appearance in the medulla. Well-localized touch makes its first synapse in medulla nuclei whose axons cross the midline and project to the thalamus. Proprioceptive body input enters the cerebellum via the inferior cerebellar peduncle.

The corticospinal tract, the most prominent motor output path, carries the output of the fifth floor motor cortex directly to motor neurons in the spinal cord. This tract of axons crosses the midline in the lower medulla and proceeds downward. Various other output tracts bound for the interneurons of the spinal cord also pass through the medulla. These tracts affect posture, balance, reflexes, and coordination and originate in the nuclei of the upper medulla, pons and midbrain. There are four cranial nerves that interface into the medulla oblongata carrying visceral input and output affecting the pharynx, larynx and tongue. An important medulla nuclei complex affecting motor output is the inferior olives. The output of the inferior olives is an important part of the cerebellum's learning neural circuit.

Much of the space in the medulla houses the massive interconnection tracts that cross the midline. The medulla oblongata provides the connection between the spinal cord and the cerebellum, thalamus, and the various motor output centers of the brain stem. Cranial nerve nuclei in the medulla support heart rate, respiration, digestion, and reflex actions such as saliva production and swallowing. It is these reflex systems that represent the medulla's greatest role. The smooth muscles of all of your glands and organs are innervated by axons from the medulla. The control of heart rate, lung respiration, intestines, liver, pancreas, digestion and a host of other housekeeping activities are all innervated and controlled through implementation of medulla oblongata neural feedback circuits.

The second brain stem room sits on top of the medulla and does for the head what the medulla does for the body. The pons contains pontine nuclei that receive input from the entire cerebral cortex and project exclusively through the middle cerebellar peduncle to the cerebellum. The entire, large, middle cerebellar peduncle is made up exclusively of pontine axons and is a major interconnection path of the brains skeletal output system. It is these massive interconnections that gives the pons its characteristic frontal bulge and clearly delineates it from the medulla. The lower nuclei of the auditory input system reside in the lower pons.

Pons nuclei receive somatosensory input from the head and provide motor output to the muscles of the face and jaw. Many semi reflexive actions such as chewing are supported by nuclei of the pons. The brain stem's ubiquitous reticular formation is located in the midline throughout the pons. A very important nucleus of the pons is the locus ceruleus. This small bilateral nucleus in the upper pons contains only approximately 30,000 neurons yet provides approximately 50% of the neurotransmitter norepinephrine to the rest of the CNS. Norepinephrine in the brain is analogous to adrenaline in the body and this small nucleus plays a major role in controlling your level of alertness and awareness. Axons from the locus ceruleus span out to the entire CNS and exert control on the entire CNS.

The functions unique to the pons include the output system pontine nuclei providing input to the cerebellum, neural support for the mastication of food, and the locus ceruleus providing neurotransmitter control over the entire CNS. The medulla is required for life to exist. The pons is required for that life to be awake and alert. Damage to the pons may not result in death but commonly results in unconsciousness or coma.

The midbrain is really the first neural structure we humans think of as a real brain. Sound input, visual input and somatosensory input are combined in the midbrain to build a neural representation of the external world. If a firecracker goes off near you, your midbrain utilizes its spatial real world map to orient your head and eyes to the location of the startling input. With the addition of the midbrain, we are almost up to the brain of a snake. The midbrain represents the top third of your reptilian brain with many input and output paths and neural nuclei that are similar to the lower pons and medulla.

There are approximately twenty million axons from the cerebral cortex that project downward on each side of the midbrain. Most of these axons are bound for the pontine nuclei in the upper pons and lower midbrain and thus on to the cerebellum. Only approximately 5% of this massive cerebral output is delivered to the motor neurons that directly drive skeletal muscles.

In humans, the highest-level functions implemented by the midbrain are its involvement in your auditory and visual systems. The inferior colliculus in the upper pons and lower midbrain receives and integrates audio input from both ears to differentiate the direction of sound in your environment. Reflex actions due to the receipt of loud noises are implemented here. The audio related output of the inferior colliculus goes to the thalamus and on to the cerebral cortex.

The superior colliculus in the upper midbrain receives visual input directly from the optic tract as well as from the visual cortex. The superior colliculus signals directly to interneurons in the upper spinal cord to effect reflex actions due to visual stimuli in the visual field. Controlling eye saccades and orienting the head and eyes to intercept a moving visual stimulus are the types of behaviors implemented by the superior colliculus. Cranial nuclei controlling your eyeballs are a part of this visual control circuitry.

The largest nuclei in the midbrain are the substantia negra. These nuclei produce the neurotransmitter dopamine that they deliver to synapses in the rest of the brain. The neurotransmitter dopamine is delivered mainly to the fourth floor basal ganglia and the fifth floor frontal cortex, areas involved in output motor control. Lack of dopamine production causes Parkinsons disease and is characterized by a lack of output muscle control.

Based on current input from the environment, the reticular formation controls what deserves attention and the level of alertness required. Input concerning pain is an especially effective ingredient in raising the level of reticular stimulation of the rest of the nervous system. Extreme input will cause the reticular formation to function like an alarm. Have you ever watched one of those nature programs where the crocodile lies motionless as the camera advances and then suddenly sprints into the water? That crocodile's reticular alarm just went off.

The reticular formation extends the entire length of the brain stem and is made up of discrete nuclei in three distinct areas. All of these reticular areas are involved in regulating the state of the entire CNS. These areas are interconnected with many nuclei in the brain stem and receive collaterals from the many neural paths transgressing the brain stem. The reticular formation monitors all types of input. If the collateral inputs to the reticular formation from sensory input axons are severed, the sensory input is ignored by higher-level brain structures. The sensory input receives zero attention. Reticular output proceeds both down to the spinal cord and up to higher brain levels. Your reticular formation provides the same functions for you as that crocodile's does for it.

The brain stem represents a complete integrated brain in lower animals. Most of the human brain stem's duties in keeping your bodily functions operating within certain constraint boundaries are unchanged from these simpler animals and these functions are essentially unconscious. Some higher-level brain stem functions, such as directing the eyes within the environment, are under the conscious influence, control and supervision of the cerebral cortex.

Up to this point, we have been discussing components of the CNS that are hard wired. These structures and their functions are defined in the gnome and expressed during growth of the embryo. In the cerebellum we encounter a different sort of brain component. The neural connections made by the neurons of the cerebellum are altered with experience. The cerebellum is a learning structure. The cerebellum allows you to learn coordinated skeletal movement.

Do you remember learning to ride a bike? Mastering balance, pedaling, and steering in a coordinated manner required a large amount of practice. Once you learn to ride a bike, you never forget how. Pattern detection learning supported by cerebellar neurons allowed you to master the bicycle and those patterns are still there for you. The cerebellum is part of your reptilian brain stem and is included in the neural make up of all reptiles.

There are three learning structures in the brain, the second floor cerebellum, the fifth floor cerebral cortex, and the fourth floor basal ganglia. The cerebellum and cerebral cortex share many structural similarities. Physically they resemble balloons. The skin of these balloons is made up of neuron cell bodies and the space inside the balloons is filled with input to and output from those neurons. The cerebellum and cerebral cortex are gray matter balloons filled with axonal white matter. The gray matter skin consists of a regular array of neurons and neural circuits. The arrangement of neurons and their interconnections are identical over the whole balloon skin. These learning structures start out as blank sheets of neural circuits that become modified by experience.

The experience that modifies the pattern detection of cerebellar neurons is control of your skeletal muscles. The cerebellum first appeared with the lower primitive brain stem and provided the ability to learn balance. Your cerebellum has evolved to add space and capability. The first evolutionary addition occurred with the growth of the upper brain stem and enabled learning output control for all motor functions provided by brainstem nuclei. The final addition evolved in conjunction with the cerebral cortex and learns complex voluntary movements controlled by the motor cortex.

The cerebellum occupies approximately 10% of your brain volume but contains over 50% of all the neurons that make up your brain. Over half of all the neurons in the human brain are located in the cerebellum and are dedicated to learning skeletal muscle coordination. The cerebellum packs an amazing number of neurons into a very small space.

Cerebellar neurons form a regular array of neural circuits that perform a standard function for each and every skeletal muscle. The cerebellum receives proprioceptive input generated by muscle spindles and Golgi tendon organs conveying the position and contraction state of every skeletal muscle. The cerebellum also receives collateral input from every axonal projection targeting motor neurons. This input comes from the vestibular nuclei in the brain stem that controls balance, the other motor output nuclei of the brain stem that control muscle tone, antigravity, and other reflexive motion, and the motor output from the cerebral motor cortex that controls voluntary movement. A final cerebellar input is a direct projection from essentially the entire cerebral cortex that conveys the state of the entire cerebral cortex.

The cerebellum receives input about what your body is doing, what it is being asked to do, and the current state of your entire cerebral cortex. This is a massive amount of input. The cerebellum has many more inputs than outputs. The cerebellum projects to the motor output structures that drive skeletal muscles. That's it. With all of that input and over half of your brain's neurons, the cerebellum is only concerned with control of your skeletal muscles. The cerebellum provides learned skeletal output patterns back to all of the neural structures driving motor output and it does this for every muscle in your body.

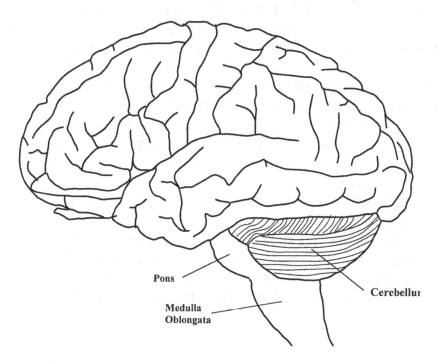

Figure 2-7: The Cerebellum

The cerebellum is located at the rear of the brain stem adjacent to the pons. (Ref. Fig. 2-7) The small bump on the lower back of your skull marks its location. The cerebellum wraps around the brain stem and, like the cerebral cortex, has a folded surface to increase the surface area in the limited space available. Unlike the cerebral cortex, the cerebellum's folds are finer and are oriented into regular long horizontal ridges and folds. It is the axonal connections within the growing cortex in the embryo that determine the shape of the folding and the axonal connections within the cerebellar cortex are extremely regular and organized horizontally.

The cerebellum grows like an expanding balloon in the embryo. When the balloon runs out of space it continues to add surface area. This further expansion causes the folds to appear, similar to stuffing this page into a tennis ball, resulting in maximum cortex surface area and thus maximum neural capacity for the available space.

The three most important neurons in the cerebellum are the purkinje neurons that perform the actual learning, vast numbers of extremely small granule neurons that pattern detect cerebellar input, and cerebellar nuclei neurons that provide the output from the cerebellum. (Ref. Fig. 2-8) The thin layer of neurons that comprises the skin of the cerebellum is made up of three distinct layers. The inner most layer, the granule layer, consists of huge numbers of granule neurons that receive input and signal to the dendritic arbors of purkinje neurons in the structured outer layer. It is the massive numbers of very small granule neurons that causes the cerebellum to contain over 50% of the neurons in your brain.

The middle purkinje layer consists entirely of purkinje neurons that are lined up in orderly rows and are the only neurons whose axons leave the cerebellar cortex to signal to cerebellar nuclei in the middle of the cerebellar balloon.

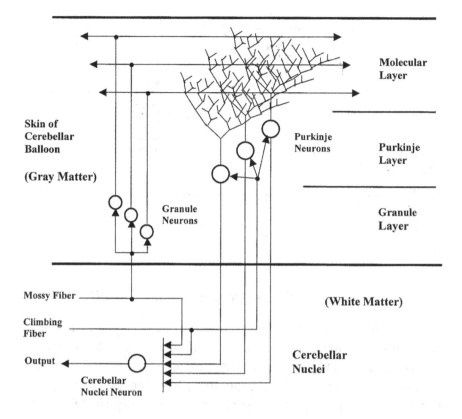

Molecular
Layer

Skin of
Cerebellar
Balloon

Purkinje
Neurons

Purkinje
Layer

(Gray Matter)

Granule
Neurons

Granule
Layer

Mossy Fiber

(White Matter)

Climbing
Fiber

Output

Cerebellar
Nuclei

Cerebellar
Nuclei Neuron

Figure 2-8: Cerebellar Layers

The outer-most layer, called the molecular layer, contains a perpendicular array of neuronal synapses. Axons from granule neurons in the inner granule layer ascend to the outer layer and run horizontally for some distance in the outer molecular layer. These axons cross and synapse with the dendrites of purkinje neurons whose cell bodies make up the middle layer. The dendrites of purkinje neurons are shaped like a sea fan and project up into the outer layer perpendicular to the horizontal axons. Picture rows of closely packed vertical sea fans with large numbers of horizontal axons running through them. This perpendicular structure of the outer molecular layer is consistent over the entire cerebellar cortex.

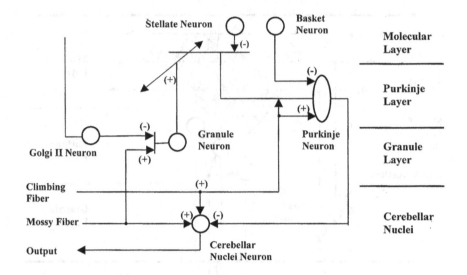

Figure 2-9: Standard Cerebellar Circuit

To understand what the cerebellum is doing, lets perform circuit analysis on a single cerebellar neural circuit (Ref. Fig. 2-9). This is circuit analysis, not science. Based on the inputs, outputs, and circuit structure, what appears to be the function of this circuit and how does it work. The circuit depicted in Figure 2-9 shows single inputs for simplicity. All of the neurons in Figure 2-9 have a dendritic tree with multiple inputs and axon collaterals which signal to multiple neurons. Similar to all neural components, the cerebellum exhibits a high degree of parallelism.

Each standard cerebellar circuit projects to a motor output area that drives a particular set of muscle cells. That motor output area is a second floor brain stem nucleus or the fifth floor motor cortex. The projection provided by the cerebellum is the learned motor output concerning that same set of muscle cells.

There are two kinds of inputs to the cerebellar circuit, the mossy fiber axon and the climbing fiber axon. The mossy fiber provides input about the muscle cell group this circuit affects. This input includes the state of the muscle cell group, motor output to the muscle cell group, and the state of your cerebral cortex. The mossy fiber signals directly to the cerebellar nuclei neuron and to granule neurons in the internal granule layer.

Mossy fibers are excitatory and terminate in numerous short collaterals that give them their name. Granule neurons receive input from many mossy fibers. The axon of the granule neuron ascends to the molecular layer and forms a long horizontal axonal appendage that synapses with many perpendicular purkinje dendrites.

Each climbing fiber input innervates ten or less purkinje neurons and a purkinje neuron receives input from only one climbing fiber. This axonal fiber attaches to the purkinje neuron cell body and dendritic arbor like ivy growing up a tree. Synaptic connections are made to both the cell body and extensively to the purkinje dendritic arbor. This input is strongly excitatory to the purkinje neuron and a climbing fiber action potential always causes the purkinje neuron to fire. The climbing fiber also sends a collateral excitatory signal to the cerebellar nuclei neuron.

The climbing fiber appears to be a control input. If it is on, it causes the purkinje neuron to fire continuously. This has the effect of putting the purkinje neuron into a learning mode. The climbing fiber causing the purkinje neuron to fire has the effect of strengthening all of the synapses that are active during that firing. Remember the Hebbian principle of how neurons learn. Synapses that are active when the neuron fires are strengthened. An active climbing fiber causes the purkinje neuron to fire and increases its sensitivity to its current dendritic pattern. The climbing fiber is very active as a new movement is learned and transitions to relatively quite after a movement is mastered. It is the learning of pattern detection by purkinje neurons that allows you to learn to control your body.

The firing purkinje neuron's strongly inhibitory output projects to the cerebellar nuclei neuron and inhibits the cerebellar nuclei neuron. This has the effect of stopping cerebellar output while the purkinje neuron is in learning mode. If the climbing fiber is off, the cerebellar circuit is free to provide output.

The purkinje neuron's dendritic arbor is shaped like a fan that is strictly aligned in the vertical plane and supports synapses with as many as 200,000 horizontal granule axons. That is a tremendously large input pattern recognized by the purkinje neuron to modulate control of one set of skeletal muscle cells. The purkinje neuron sends an inhibitory signal to the cerebellar nuclei neuron modulating that neuron's output through inhibition.

There are three other inhibitory neurons that are a part of every cerebellar circuit. Golgi II neurons are inhibitory neurons located in the granule layer that have three-dimensional dendritic arbors in the molecular layer and serve to regulate granule neurons. The purkinje neuron also receives input from two inhibitory neurons in the molecular layer. Stellate neurons dampen granular input on the purkinje's dendritic arbor and basket neurons send strong inhibitory input to the cell body of the purkinje neuron. The basket neurons detect when their row of purkinje neurons are firing due to pattern detection of granule input and shut down other adjacent purkinje neurons. These types of winner take all neural circuits have the effect of disabling all but the strongest comparing neural circuits and filtering out extraneous noise. This example of winner take all inhibition is a mainstay of brain neural makeup.

There is only one output from the cerebellar circuit. That output comes from the cerebellar nuclei neuron in the middle of the cerebellum balloon that projects back to an output structure driving the skeletal muscle cell group this circuit serves. Since there are multiple output structures driving each skeletal muscle, there are multiple cerebellar circuits providing input about each skeletal muscle, one for each output type.

Lets recount the results of our circuit analysis. It appears that the cerebellar circuit transitions from output mode to learning mode under the control of the climbing fiber input. An active climbing fiber puts the cerebellar circuit into learn mode. In learn mode the purkinje neuron fires repeatably inhibiting the cerebellar nuclei neuron from providing output and causing its dendritic arbor synapses to become more sensitive to the current input pattern. An inactive climbing fiber allows the cerebellar circuit to be in output mode. The cerebellar nuclei neuron pattern detects input concerning the skeletal muscle cells it supports and fires accordingly unless inhibited by the purkinje neuron. The purkinje neuron is pattern detecting a very large amount of input concerning these skeletal muscle cells and quite possibly the state of the body area surrounding these muscle cells. Lets remember that this is circuit analysis and not science. This is how the cerebellar circuit appears to work.

The cerebellum contains three distinct areas that perform an analogous function for three different neural output centers. The oldest of these areas, located in the bottom of the cerebellum, is called the archicerebellum and interacts with vestibular nuclei in the brain stem to control balance and equilibrium during head and body movement. The purkinje neurons in this area project their axons directly to the vestibular nuclei, the only area where cerebellar output comes from other than the cerebellar nuclei. This area of the cerebellum can be observed in action as a baby learns to walk.

The next area of the cerebellum to evolve is called the paleocerebellum. This area is located at the top of the cerebellum and sends its output to motor output nuclei in the brain stem. Proprioceptive input from the entire body is received here and cerebellar nuclei provide input to brain stem nuclei providing unconscious skeletal control.

The most recent addition to the cerebellum, the neo-cerebellum, occupies the large middle portion of the cerebellum and interacts with the motor cortex to control learned voluntary movement. Neocerebellar input is received from pontine nuclei in the pons and lower midbrain that relay output from the entire cerebral cortex. Neocerebellar nuclei project to the motor output portion of the cerebral cortex through the thalamus. This circuit must be operational if you want to learn to play the piano. The cerebellum is not a skeletal muscle output structure. It provides learned output patterns to output structures in order to effect control.

The cerebellum does not provide a hardwired response. It is at birth a relatively blank sheet that learns to control your output movements as you grow, mature, and learn new tasks. The cerebellum has grown in capability as brains have evolved. These systems evolved in series and each successive cerebellar system does not replace but is built on top of and compliments the existing system.

Without an archicerebellum you have trouble standing, without a paleocerebellum you have trouble walking, and without a neocerebellum you cannot play the piano. The jump from walking to playing a piano is large and is indicative of the importance of the latest addition to your cerebellum.

Floor #3 - The Hypothalamus (Limbic System) and Thalamus

The third floor of your brain building contains only two rooms, the hypothalamus and the thalamus. The small, very old hypothalamus is the output command center of your limbic system and controls the hormonal state of your body. The limbic system is made up of a number of brain structures, including the oldest portions of your cerebral cortex, which combine to produce your emotional state. The thalamus evolved in conjunction with the cerebral cortex and consists of a collection of nuclei that interconnect with all areas of the cerebral cortex. The most obvious function of the thalamus is to gate all incoming sensory information to the cerebral cortex.

Emotions and physical sensations such as hunger and thirst are produced by the hypothalamus and a collection of brain structures called the limbic system. The limbic system contains both hardwired components and very old portions of the cerebral cortex that support learning. This ancient emotional brain system has existed for a very long time. The human brain, with its greatly expanded cerebral cortex, is the new kid on the block and this old emotional system continues to have tremendous influence over your behavior.

Located just below the thalamus, the hypothalamus comprises less than one percent of the brain. This small size is not indicative of its importance or the complexity of its functions. The hypothalamus produces hormones in order to maintain your entire body in a consistent state. They are produced in response to hunger, light, sexual stimulation, stress and a myriad of other external inputs in order to assist your body in dealing with those inputs in a coordinated way. Hypothalamus output includes axonal connections to two glands that control the production of hormones. The pituitary gland is called the master gland of the body and lies just below the forward part of the hypothalamus and the pineal gland lies below and to the rear.

The hypothalamus is extensively interconnected with all of the brain structures that make up the limbic system and contains the output portion of feedback control circuits that regulate the state of your body. Through projections to the brain stem and spinal cord, the hypothalamus exerts control over the autonomic nervous system. The hypothalamus also projects to the reticular formation in the brain stem and has an impact on your state of neural alertness. The input to and control functions of the hypothalamus are not directly available to the cortex and its functions are a part of your autonomic, unconscious nervous system.

The limbic system lies between your reptilian brain stem and your more recently evolved cortical areas. This highly interconnected emotional neural system projects to the frontal cortex via the thalamus and receives projections from the frontal cortex. The frontal cortex is intimately involved with the limbic system but is not considered a part of the limbic system. The frontal cortex does not produce your emotional state. These interconnections with the frontal cortex allow you to be conscious of your emotions and affect some conscious control over those emotions. Think happy thoughts.

It is through the actions of the limbic system that you acquire feelings about your environment and the state of your body. Low blood sugar elicits a feeling of hunger that affects your mood and your behavior. Stress, loss, separation, and conflict are examples of your frontal cortex interacting with the limbic system to affect your emotional feelings. The limbic system determines whether you feel happy, sad, exited, apprehensive and all of the other descriptive words you use to describe your complex emotional state. The emotional state produced by the limbic system motivates you to behave in certain ways.

Your limbic system produces your emotional state and your emotional state affects the strength of memory formation, your ability to learn and how you behave. How you feel has a profound impact on all of the higher-level mental capabilities that characterize the human brain. With the newly developed cerebral cortex and corresponding conscious logical control over behavior, humans believe that they can control their emotions. That conscious control operates on top of a very old and powerful neural system that drives coordinated behavior through emotions and motivation. Hunger, thirst, anger, and fear are but a few of the many and varied set of emotions that motivate humans to behave in certain programmed responses. Conscious control of behavior influences and operates in conjunction with this older neural system.

The thalamus is a hardwired collection of nuclei that evolved with and are literally tied to the cerebral cortex. The thalamus forms one end of a common neural circuit with the cerebral cortex called thalamocortical loops. A great deal of the white matter filling the inside of your head is made up of these interconnections. These bi-directional neural circuits completely interconnect the thalamus with the entire cerebral cortex.

The thalamus is a sausage shaped structure just over one inch long that is located medially at the centerline of the brain. The combination of the left and right thalamic structures sit in the center of the overlying cerebral cortices surrounded by the components of the limbic system.

There are two kinds of thalamic nuclei, gating type nuclei and reticular type nuclei. Most of the nuclei that make up the thalamus are gating type nuclei. Their axonal projections target a particular cerebral destination and they receive reciprocal connections from that same cerebral destination forming a loop, a thalamocortical loop. There are two reticular type nuclei, the intralaminar nucleus lies between the gating nuclei and the nucleus reticularis forms a thin sheet like covering over the thalamus. The axonal projections of the intralaminar nucleus disperse widely over the cerebral cortex and are involved in neural arousal and attention. The axonal projections of the sheet like nucleus reticularis provide inhibiting input to gating type neurons and have inhibitory control of thalamocortical loops.

The thalamus acts like an input neural switchboard between the cerebral cortex and the rest of the central nervous system. With minor exception, all neural signals bound for the cerebral cortex pass through and are gated by the thalamus. These cerebral inputs include sensory input that is comprised of all of the senses, motor control input from the cerebellum, and limbic system input concerning emotional status.

The thalamus is much more than just an input switchboard. The thalamus has reciprocal connections with the entire cerebral cortex, including association regions of the cerebral cortex that receive no gated subcortical neural input. The thalamus receives approximately ten times more input from the cerebral cortex than the thalamus sends to the cerebral cortex.

Reciprocal connections with the cortex allow incoming neural input to be modulated. Sensory input signal gating through inhibition is common for all stages of input and is especially effective in the thalamus. Thalamic gating neurons receive sensory input and reciprocal input from the cerebral cortex and fire to pass on receipt of sensory information when not inhibited. Various inputs, including reticular input, influence the responsiveness of a particular gating thalamic neuron. In this fashion the amount and type of sensory input passed to the cerebral cortex is modulated.

The thalamus is comprised of many nuclei. Each individual gating type nuclei interconnects with one area of the cerebral cortex and, if it gates input to the cerebral cortex, handles only one kind of cerebral input. The names of thalamic nuclei are a concatenation of their position within the thalamus. The thalamus is divided into medial (middle), lateral (side), ventral (bottom), dorsal (top), posterior (rear), and anterior (frontal) areas. (Reference Figure 2-10) The ventral posterior medial and lateral nuclei receive somatosomatic input from the head and body and project to the primary sensory cortex. The lateral and medial geniculate nuclei gate visual and auditory input to the primary visual cortex and primary auditory cortex respectively. Limbic input is gated by the anterior nucleus to limbic cortex and by the dorsomedial nucleus to the frontal cortex. Motor input from the cerebellum is gated by the ventral lateral nucleus to the motor cortex. The pulvinar nucleus located in the posterior portion of the thalamus provides connections to visual association cortices located in the parietal, temporal and occipital lobes.

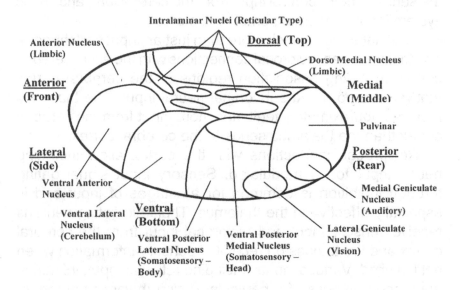

Figure 2-10: Thalamic Nuclei

The natural inclination is to compare the thalamus to a simple input switchboard for the cerebral cortex and it indeed performs that function. The problem with this comparison is both the word simple and the fact that the relationship between the thalamus and the cerebral cortex involves more than just sensory input. The thalamus forms thalamocortical loops with every area of the cerebral cortex, not just sensory input areas. This neural circuit performs a standard function for each and every area of the cerebral cortex. Understanding that standard function is a necessary prerequisite to understanding how your brain enables human intelligence.

Floor #5 - The Cerebral Cortex (The Top Floor)

Because our brain building tour is tracking the neural path of sensory input to neural output, we are going to skip the output related fourth floor and go right to the top, the cerebral cortex. The fifth floor is different from the rest of the brain building floors. The fifth floor is one very large room with no walls and no offices. It resembles a very large open office floor filled with cubicles. This large office space is divided into areas of cubicles that perform specific functions, just like a real office.

The human cerebral cortex is the ultimate learning structure in the animal kingdom. Everything you have experienced and learned, the sum total of your life that results in behavior patterns that define you, are stored in your cerebral cortex. The cerebral cortex is not required for life to exist but is required for intelligence to exist. Compared with other mammals, you have an amazing amount of cerebral cortex crammed inside your skull. You have so much that it folds into hills (gyri) and valleys (sulci) as it grows and overfills the cranial vault during fetal development.

There are approximately one hundred million neurons per square inch and approximately fifteen billion neurons total in the human cerebral cortex. Those two approximations translate into a human cerebral cortex that is just over one square foot in area. While the cerebral cortex does not contain as many neurons as the cerebellum, it does contain a very large number of neurons.

Like the cerebellum, the cerebral cortex is a thin-skinned sheet of neurons resembling a balloon filled with white matter. The white matter consists of the myelinated axons that interconnect the various areas of the cortical sheet and connect the cortical sheet with lower brain structures. The cerebral cortex is divided into five major lobes, the emotional system limbic lobe, the vision occipital lobe, the environmental space parietal lobe, the language and object recognition temporal lobe and the behavioral frontal lobe. The most recent additions to your cerebral cortex constitute your neocortex and contain a six-layer arrangement of neurons in a remarkably regular structure throughout.

Your large amount of cerebral cortex is what allows you to be intelligent. Frogs have no neocortex. Fish and all mammals do have some amount of neocortex but in paltry amounts when compared with humans. Understanding the function and operation of the human cerebral cortex is a prerequisite to understanding human intelligence. We will first cover the oldest portion of your cerebral cortex, the hippocampus. The hippocampus performs a distinct and separate function and a functioning hippocampus is necessary for the normal operation of the rest of the cerebral cortex.

The hippocampus was the first area of cerebral cortex to appear. Its neural circuitry is different from the rest of the cerebral cortex and it is the only portion that is not reciprocally connected with the thalamus via thalamocortical loops. The hippocampus is required for new patterns to be learned and subsequently recognized in the rest of the cerebral cortex. Your hippocampus is required for the proper operation of memory. The cerebral cortex, the ultimate neural learning structure, cannot learn without a functioning hippocampus.

The most revealing evidence regarding hippocampal function is the famous case study of a patient called HM. HM had severe epilepsy and in an attempt to alleviate his symptoms, a surgeon removed his hippocampus and some surrounding supporting cortex and nuclei on both sides of his brain. The operation was a success as far as the epilepsy was concerned but had some unfortunate side effects for the patient.

HM from that day forward was unable to form new memories. His intelligence seems unchanged and he remembers everything that happened to him prior to the operation. He remembers nothing that has happened to him since his operation. He appears normal and his short-term memory seems unaffected but he actually has complete amnesia for all things discarded from his immediate attention. If you work with HM, he will not recognize you when you meet in the morning and you must reintroduce yourself daily. It turns out that HM is able to learn motor output tasks. If he is trained to perform some intricate mechanical task, his performance will continually improve with training but he will not recall the training. This case and other similar cases give us clues as to the function of the hippocampus.

The hippocampus is referred to as archeocortex and its function and physiology are common in all animals that have a cerebral cortex. Unlike the more modern six-layer structured neocortex, the hippocampus exhibits a three-layer structure. In many ways the hippocampus more closely resembles the cerebellar neural circuit structure than modern neocortex. Like the cerebellum and the rest of the cerebral cortex, the hippocampus is a learning structure and is not hardwired.

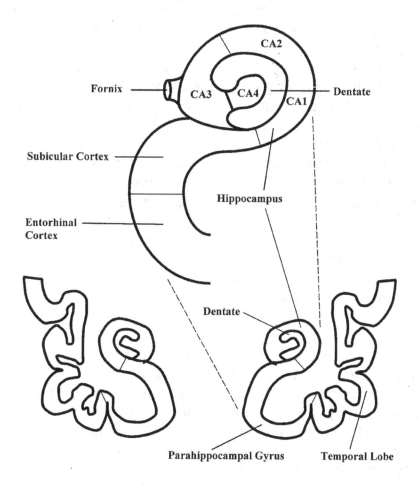

Figure 2-11: Hippocampus/Dentate

The cerebral cortex resembles two interconnected balloons on either side of the skull. The wide mouth area of this balloon like structure forms the limbic lobe of which the hippocampus is a part. The hippocampus lies on the medial, lower side of this balloon opening and is rolled up into a loop that is elongated from front to back. (Ref. Fig. 2-11) The hippocampus fold is arbitrarily divided into four sections, CA1 through CA4. Inside of this tubular fold is a neural layer called the dentate. The sheet of neurons that comprise the skin of your cerebral balloon transitions from the hippocampus to subicular cortex and then to entorhinal cortex. These cortical areas jointly comprise the parahippocampal cortex that transitions into the neocortical temporal lobe.

The hippocampus is reciprocally connected with all neocortical areas through parahippocampal cortex that serves as a transition zone between the six-layer neocortex and the three-layer hippocampal archeocortex. This interconnection is called the alveus and is the first stage of the communication path between the hippocampus and the association areas of the entire neocortex. A second communication path is the fornix that connects the hippocampus with the rest of the limbic system.

The neural structures that interact with the hippocampus are the higher-level association areas of the cerebral cortex and the limbic system. Higher-level cortical association areas build more complex patterns from primary input representing objects, space, and behavior and these areas have extensive reciprocal communication with the hippocampus. Most areas of the cortex reciprocally communicate with the hippocampus through parahippocampal cortex at some level. The limbic system interacts with the hippocampus through the fornix and more indirectly through the rest of the limbic lobe.

Since the famous case of HM, everyone has known that the hippocampus is involved in the formation of memories. The issue is what memory function does it perform. For a memory system to work there are three functions that must be operational. The memory has to be written or caused to be stored, there has to be a storage location, and the memory must be able to be read or accessed. Since HM has all of his memories intact from prior to the removal of his hippocampus, the hippocampus must not be the storage location nor does it provide the read function. That leaves some form of the write function for long-term memories as the hippocampal contribution.

A human memory is a multidimensional web of associations. If you see an apple, your brain recognizes the pattern of disparate edges, lines, and color as a known object. You immediately know the written and spoken word associated with it, perhaps in multiple languages, as well as its taste and smell. These associative links that tie together the visual pattern into an object and tie together other portions of the cortex that store sound patterns and words have been built over time. HM recognizes an apple and knows these things about an apple because he has previously stored the associations required for that recognition. However, if he re-encounters a fruit new to him since his operation, he does not recognize it. He is unable to write, store, or read the associations necessary to tie these things together.

The structure of the hippocampus consists of just three layers. (Ref. Fig. 2-12) The hippocampus actually contains two neural structures, the three layer folded hippocampus and the dentate nucleus that lies inside this fold. Much of the neural input to the hippocampus enters through the dentate nucleus and is pattern detected by the dentate granule neurons. The neural output of the dentate nucleus is so extensive that the white matter causes the neuron layers to be separated into two distinct structures. These dentate granular neurons project only to the hippocampus proper.

We will perform circuit analysis of the standard hippocampal neural circuit depicted in figure 2-12 to determine what the structure and neural circuits appear to be accomplishing. The hippocampus provides the interconnection of two major brain systems, the limbic system and the cerebral cortex. The hippocampus interconnects your old reptilian emotional brain with your new cerebral cortex areas where memory patterns are stored.

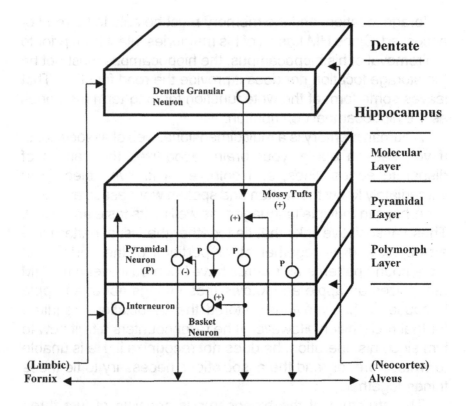

Figure 2-12: Hippocampus/Dentate Neural Wiring

Every portion of the hippocampus is dedicated to integrating both limbic and cortical patterns. Input from both the limbic system and cerebral cortex enter the polymorph layer and dentate structure. The dentate is the main input structure of the hippocampus and contains a very large number of granule neurons. Interneurons and granule neurons pattern detect this input and project to the dendritic arbors of pyramidal neurons in the molecular layer. These molecular layer interconnections are called mossy tufts reminiscent of the cerebellum. Basket neurons in the polymorph layer receive collateral input from pyramidal output neurons and inhibit adjacent pyramidal neurons.

Pyramidal neurons perform second level pattern detection of limbic and cortical input and are the only neurons that provide output from the hippocampus. That output is provided back to both the limbic system and cerebral cortex. The hippocampus sits between two neural systems that signal "this is what I'm feeling" and "this is what I'm observing". The pattern-detected combination of those two inputs is provided as feedback to both neural systems.

That the hippocampus is required for the ongoing formation of memories has been accepted for many years. The actual memory related function that the hippocampus plays is still a matter of debate. Loss of the hippocampus has no effect on short-term memory. The brain is quite capable of dealing with the complex world in real time without a hippocampus. A new fruit with a new name, taste and smell can be set on a table and discussed with HM for a great length of time. The relevant facts about this fruit will not be lost until attention is diverted to something else. Once attention is diverted however, all of the associations that allowed the fruit to be recognized, named, and discussed are lost. HM can still remember everything that happened prior to losing his hippocampus.

Thus the hippocampus is not the storage location of the patterns that comprise memory or is it required to access those stored memory patterns. HM, without his hippocampus, either cannot store new memory patterns or cannot store them in a way that allows them to be recognized.

The hippocampal cubicle area is unique but the rest of the fifth floor cerebral cortex is exactly the same. A vast sea of identical cubicles fills the entire room. There are no walls, no offices, no conference rooms, and no CEO or CFO space. There are no managers of any kind. This looks like the ultimate in office democracy.

There are four separate pairs of elevators that bring input to the fifth floor and each opens into a different cubicle area where workers handle the input workload. The elevators bringing left and right visual input open in the back of the room. The audio elevators open on the sides of the room. The skin and taste input elevators open in the middle of the room and large freight sized elevators bringing emotional limbic input open in the front of the room. Input concerning smell enters via an old staircase.

Each area of cubicles pattern detects their input and passes their output onto adjacent areas, which perform their pattern detection and pass work on again. Each cubicle area also sends its pattern recognition results back to the cubicle areas that provide input to it. Areas of cubicles away from the input elevators handle more abstract work and are called association areas. Some cubicle areas pass work to non-adjacent areas of the fifth floor. A vast array of wires underneath the fifth floor enables the passing of work between cubicles. Every cubicle area has security cameras like in a bank. TV monitors on the second, third and fourth floors allow the entire fifth floor to be watched continuously at all times. Those lower floors are watching at all times and what they see determines what they do.

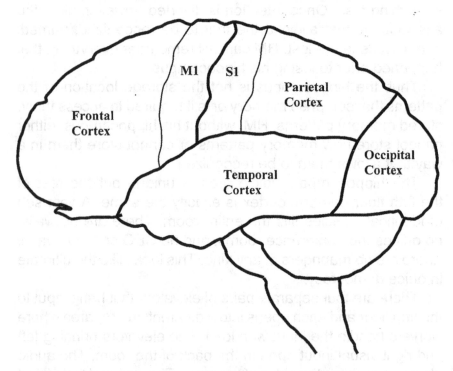

Figure 2-13: Cerebral Cortex

The cerebral cortex is what you see when you look at a picture of a human brain. (Ref. Fig. 2-13) You see the familiar fissures that divide the brain into left and right hemispheres and the edges of many folds. The exposed gray skin represents a portion of the entire cortex as most of the surface area is buried in the fissures and valleys of folds. This gray skin is a continuous sheet of neurons with no holes or gaps that grows in the embryo as a balloon fills with air. It grows from the oldest parts of the cortex, the open mouth of the balloon lower limbic collar, to the new fast growing neocortex that comprises most of the human cerebral cortex.

At birth this cortical sheet is fully populated with neurons but incompletely interconnected. All neocortex consists of six distinct layers of neurons, its thickness and structure consistent throughout. The number of neurons in the various layers varies depending on the function of a particular section of cortex but the number of neurons in a square millimeter of cortex remains the same at approximately 150,000.

The inside of your cortical balloons contain the white matter from the input and output signals of the cerebral cortex. Reciprocal thalamocortical loop projections between the thalamus and the cortex are called the corona radiation. Reciprocal projections to the hippocampus and projections to the basal ganglia and beyond are also a part of this radially shaped collection of axons. Reciprocal communication between areas of the cerebral cortex comprises a large portion of your white matter. Reciprocal ipsilateral connections are both short and long with major bundles of axons interconnecting the parietal and temporal lobes with the frontal lobe. Reciprocal contralateral communication literally ties the left brain to the right brain at all levels and consists of approximately three hundred million individual axons. This left and right cerebral interconnection is called the corpus callosum.

The limbic lobe is made up of a band of cortex that forms a collar like structure at the mouth of the cortex and lateral insular cortex buried between the parietal and temporal lobes. (Ref. Fig. 2-14) The hippocampus is located at the bottom of the limbic lobe. The upper portion of the limbic lobe is the cingulate gyrus. The mouths of your two balloons open at an approximate 45-degree angle towards each other.

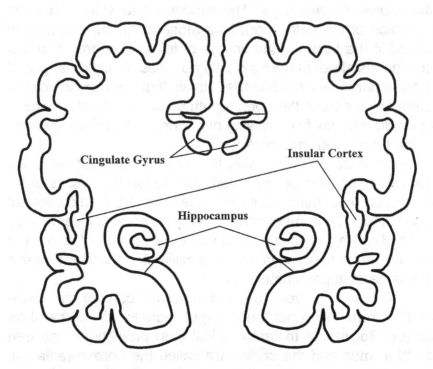

Figure 2-14: Limbic Lobe

The cingulate gyrus is a large strip of cortex that arcs from the back to the front of the brain just over the corpus callosum. It is located at the very bottom of the fissure that divides the two halves of the brain. This area plays a role in the portion of the limbic system involved in memory formation. The insular cortex, located in the medial wall of the lateral sulcus, is not visible when viewing the surface of the brain. This cortex grows slowly and is completely buried by the faster growing neocortex. The insular cortex receives input from the gustatory sense but is not primary gustatory cortex and has connections with the cingulate gyrus and the basal ganglia.

The occipital lobe located at the back of the head is concerned with vision and vision only. The primary visual cortex, V1, is located at the very posterior of the cerebral cortex and receives a visual image from the retina of the eye through the thalamus. The primary visual cortex breaks the entire image into small areas and performs pattern detection on each separate area to signal the occurrence of lines, edges, colors, and motion within each of these areas. Secondary visual areas surround V1 in concentric rings and perform ever more complex pattern detection from this input to build neural representations of what you see.

The parietal lobe contains your primary somatosensory cortex, S1, which receives input from the entire body. S1 lies in a strip at the very front of the parietal lobe separated from the primary motor cortex and frontal lobe by the lateral sulcus. Secondary somatosensory cortices lie in successively posterior bands of cortex. Most of the large parietal lobe is association cortex that combines pattern detected visual input with other sensory input to build a neural representation of the world around you. The primary gustatory cortex is also contained within the parietal lobe in close conjunction with the S1 cortex that receives input from the mouth and tongue.

The temporal lobe contains your primary auditory cortex, A1. A1 is organized in a linear fashion by the frequency of sound and performs the analogous function of all primary sensory cortexes in breaking the primary auditory input into its component parts. In humans, these auditory component parts are further pattern detected into language. The temporal lobe, especially the left temporal lobe, is where you store the patterns that comprise words. The association areas of the temporal lobe also receive visual input from the occipital lobe utilized to visually identify objects. The neurons that are involved in visual object recognition make up a large percentage of the posterior portion of the temporal lobe. The temporal lobe projects heavily to the lower parietal lobe where object recognition and spatial recognition are combined into a complete neural representation of your environment.

The large association area of the frontal cortex is the most recent expansion of your neocortex and achieves its greatest size in humans. The human frontal lobe makes up approximately one third of the entire cerebral cortex surface area and is responsible for many of the functions that make humans intelligent and unique. High-level abstraction, planning, and behavior are a few of the functions of the frontal cortex.

The frontal cortex contains a strip of motor cortex that is your primary motor output area, M1. M1 contains large output pyramidal neurons whose axons terminate on motor neurons directly driving muscle contraction. M1 lies adjacent to your primary sensory cortex S1 and is roughly a mirror image of it. An area of the frontal cortex forward of the motor control area for the mouth and tongue is critical for the production of language. The frontal cortex is also intimately involved with the limbic system. Limbic system output is projected to the frontal cortex through the thalamus and the frontal cortex affords you some conscious control over your emotions.

At birth the cerebral cortex is close to a blank slate. Areas in which pattern detection has been initiated are mainly in the areas of primary sensory input. For example, in the growing embryo the retinas of the eyes fire random continuous bursts of input that cause the neurons in the thalamus and primary visual cortex to self organize into the pattern detection areas seen in the adult. Cortical wiring that is the most efficient at pattern detecting this input is strengthened. Less efficient cortical wiring is eliminated. In this manner the pattern detection required to receive the input is built within the cortex.

This ability to alter its neural wiring to store new patterns is termed the plasticity of the cerebral cortex and is an inherent part of its nature. This plasticity continues throughout a human life. The cerebral cortex is where you build pattern detection related to object recognition, spatial recognition, language, complex associations, and behavioral responses. We will discuss the various attributes of each of the cortical areas as if they are disjoint. They are not. The cerebral cortex is one neural component and it operates as a whole.

The physical organization of the cerebral cortex is optimum for a massively parallel device and that is exactly what it is. The learning components are in a wide, thin sheet with very large numbers of parallel connections. The neural make up of this sheet of neurons is homogeneous. With the exception of the very old hippocampus, the cortex consists of six layers of neurons that are not sharply bounded or physically separated in any way but give us a context within which to describe the structure and function of the cortex. (Ref. Fig. 2-15)

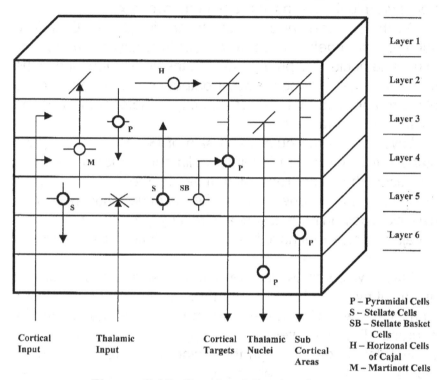

Figure 2-15: Cerebral Cortex Layers

There are two kinds of neurons that make up the majority of the cerebral cortex, stellate and pyramidal neurons. Stellate neurons receive input from the thalamus and are mostly contained in layer four. Pyramidal neurons are contained mostly in layers two, three, five and six. Pyramidal neurons are the learning type neurons and provide all output from the cerebral cortex.

Each cubicle area reciprocally interconnects with the thalamus, other areas of cortex, and supports non-reciprocal projections to subcortical areas. Interconnections with other cortical areas are supported in layers two and three. Layer two receives slightly more input and layer three provides slightly more output with other cortical areas. Layers four and six support reciprocal connections with the thalamus called thalamocortical loops. Thalamic input is received in layer four by stellate neurons and output back to the thalamus comes from pyramidal neurons in the bottom layer six.

All thalamocortical connections are ipsilateral. Output to sub cortical targets comes from layer five. Layer five projects to pontine nuclei that provide input to the cerebellum and to the fourth floor basal ganglia. Motor output control from layer five of the motor cortex M1 projects directly to motor neurons in the spinal cord or brain stem.

Layer one is the outer most skin of the cortex. This layer starts out at birth with a full compliment of mature neurons but loses those neurons over time. In the fully developed human brain, layer one contains the top of dendritic trees from pyramidal cells in lower layers and relatively few horizontal cells of cajal neurons that have horizontal dendritic trees and axonal projections.

Layer two contains many relatively small neurons consisting mostly of pyramidal cells. Layer three is close in neuron types with layer two with larger and more numerous pyramidal neurons. The dendrites of these pyramidal neurons spread through layers one through three where they receive cortical inputs from contralateral and ipsilateral cortex as well as horizontal collateral projections. Layer three pyramidal axonal projections target external cortical targets and send collaterals horizontally.

Layer four is the main target of thalamic input. Cortical areas that receive a large amount of input from the thalamus, primary sensory input areas, have a large robust layer four and are called granular cortex. Cortical areas that receive little input from the thalamus, association areas, have a shrunken layer four and are termed dysgranular cortex. Layer four contains mainly stellate neurons that receive thalamic input, stellate basket cells that are inhibitory and some small pyramidal cells. A typical stellate neuron in a primary sensory input area may receive as many as 60,000 synapses from thalamocortical projections.

Pyramidal neurons in layer five project axons to subcortical neural structures. Large pyramidal neurons in M1 called Betz cells project to motor neurons in the spinal cord and brain stem. Widespread cortical projections to the basal ganglia and pontine nuclei of the brain stem emanate from layer five. These pyramidal neurons typically have horizontal axonal collaterals that extend within the cortical layer and serve as inputs to stellate basket neurons that inhibit other pyramidal neurons.

Layer six is also a pyramidal layer that projects to the thalamus to complete the thalamocortical reciprocal connections made with that structure. The dendritic arbors of these pyramidal neurons extend up vertically to all layers.

Pyramidal neurons are the dominant cell type in the thin neuron packed sheet of the cerebral cortex in all layers except layer four. The neuron makeup of the cortex of the rat contains approximately 92% pyramidal neurons in non-layer four cortex. Pyramidal cells are oriented in a vertical fashion and have large triangular cell bodies that give them their name. The pyramidal dendrites ascend vertically as far as layer one with very large complex arbors in layers above the cell body. Pyramidal axons descend vertically with horizontal axonal collaterals. Pyramidal neurons provide all external communication projections of the cortex.

Stellate neurons are the dominant cell type of layer four and exist in all cortical layers except layer one. In granular layer four of the rat cortex, approximately 87% of the neurons are stellate cells. The dendritic arbors of stellate cells are large with intricate interconnection patterns that are very localized around the cell body. Stellate axons project in a vertical fashion both up and down to terminate on the dendritic arbors of pyramidal neurons.

Small numbers of martinotti cells exist in all layers except layer one and make up less than 1% of the cortex of rats. The dendrites of these neurons are localized and their axons ascend to layer one where they target the dendritic arbors of pyramidal cells. Stellate basket neurons are found in all layers except layer one, being more concentrated around layer four. The axonal projections of basket cells are inhibitory and surround and terminate on the cell bodies of pyramidal cells. These inhibitory cells are part of the winner take all portion of cerebral cortex pattern detection.

Understanding how the cerebral cortex communicates to other neural subsystems and also how it communicates within itself is extremely important. Communication paths within the cortex tend to be horizontal in the upper layers and vertical in the lower areas. This communication structure is horizontally dominant in layer one and transitions to almost completely vertical in layer six.

Communication between cortical areas, concentrated in layers two and three, is the major source of reciprocal connections in the entire cortex. Ipsilateral communication connects the entire cortical hemisphere together and contralateral reciprocal connections between the hemispheres interconnect the two hemispheres via the corpus callosum. These corticocortical paths also interconnect the association areas of the cortex with the parahippocampal cortex and thus the hippocampus.

Sensory input is received in specific primary sensory cortex areas and causes that cortex to have a robust layer four filled with receiving stellate neurons. Sensory input from the thalamus not only causes a robust granular layer four, it also causes granular cortex to form into highly structured assemblies of columns that extend through the entire cortex from layer one to six. These sensory input granular columns are closely packed within a matrix of dysgranular cortex. Thalamic communication occurs within the interior of the columns and inter cortical communication is contained within the dysgranular matrix. This array of columns is typical of all primary sensory input cortex.

Humans are more intelligent than other mammals, at least by our parochial definition. Discounting brain areas that grow proportionately with body size, the cerebral cortex is the brain component that is proportionately larger in humans. It follows that the cerebral cortex is what enables intelligence. That indeed is the case. To follow this logic further, the main difference between the cortex of humans and other mammals is the large amount of association cortex in humans. Therefore association cortex must be what enables intelligence. Again, that indeed is the case. Intelligence is the ability to build complex associations. Mathematics and language are the two human abilities that most clearly demonstrate your ability to abstract the world into symbols and then manipulate those symbols. These capabilities are clearly enabled by association cortex.

At birth the cerebral cortex is virtually a blank sheet that is ready to begin the job of recording your life. The cortex records your life by modifying its synaptic wiring to enhance and make accessible the neural patterns that create your existence. The cortex continues to add and alter patterns that define your perception of the world and every object in it, including yourself, throughout your life.

Floor #4 – The Basal Ganglia (The Shipping Control Department)

The fourth floor of your brain building, the basal ganglia, is the building's shipping control department. It controls your brain's output or response to current input. When we think of output, we usually think of skeletal motor output and movement. In humans, output is much more than motor control. We are talking about the whole range of actions we call human behavior. The basal ganglia are involved in the control of human behavior. The basal ganglia form a learning structure that is made up of a collection of nuclei that provide control of both motor output that is observable and human behavior that is often not observable.

Your shipping control department determines what to ship by watching what is going on up on the fifth floor. Remember those TV cameras on the fifth floor. The fourth floor watches the image of the entire fifth floor constantly in order to decide what to output. The image of the fifth floor is the only pattern input to the fourth floor. The output of the fourth floor targets the gating nuclei on the third floor thalamus that interconnect with the entire fifth floor frontal cortex. The frontal cortex, including the motor cortex M1, is the actual shipping department.

So how do you learn what to ship? Or restated, how do you learn what behavior to use? One of the major factors determining behavior selection is based on the feedback or results of that behavior. Actions that have a good outcome are usually repeated. Actions that do not have a good outcome tend to not be repeated.

Changing behavior based on the results of that behavior presents an obvious problem. That problem has to do with the time lag between a behavior output choice and the determination of the result of that behavior. Neurons perform pattern detection in real time based on current input. If behavioral learning is the modification of output patterns based on the positive or negative result of the output, how do output neural patterns get changed when the result is not known for seconds, minutes or longer? The neural patterns that caused the output to be selected are long gone. Understanding the brain's solution to this time lag problem requires an understanding of the action of the neurotransmitter dopamine. Dopamine is released into the synapses of the input portion of the basal ganglia by the axons of neurons in the substantia negra located in the brain stem. Dopamine is required for behavioral learning.

The release of dopamine is tied to the outcome of behavioral events. Dopamine release is increased by the receipt of rewards and decreased when expected rewards do not happen. Dopamine release has been monitored in animals being trained to perform a task. The animals learn to perform a task in response to a trigger signal in order to receive a reward. In the initial training, large increases of dopamine occur with the receipt of the reward. As the animal learns, the increase in dopamine occurs when the trigger occurs. The dopamine increase transfers from the reward to the expectation of a reward with training. After training is complete, the levels of dopamine release return to normal. If the expected reward is not given, a large decrease in dopamine output occurs when the reward was expected.

Lets explore a hypothetical learning situation in order to examine how dopamine facilitates behavioral learning. It is spring and you are walking across country looking for food. You come across a green bush, notice that it is covered with berries, and eat the berries. Your substantia negra releases a large positive spike of dopamine as you eat the berries. This is positive feedback. As you walk further, you notice another green bush in the distance. Green bush equals berries that taste good. You basal ganglia receives another positive spike of dopamine when you see the green bush. This positive feedback is in anticipation of a positive outcome. You proceed to the bush, eat the berries, and all is well.

You now notice that the green bushes with the berries are located in canyons with large trees. When you see a canyon with large trees, your basal ganglia receive another large positive spike of dopamine. You have now learned to seek out canyons with trees in order to find green bushes filled with berries. As you use this learned behavior, no more positive spikes of dopamine occur as you find more berries. This behavior has been learned and additional dopamine is not required. The action of dopamine caused you to learn a behavior based on the anticipation of an outcome.

Now it is late summer and you again walk across country looking for food. You notice a canyon with large trees and enter it finding the green bushes. However, the bushes have no berries. They are all dried up. Your substantia negra gives your basal ganglia a large negative spike in dopamine. You expected berries and finding none is definitely a bad outcome. After a few more detours to green berry bushes and finding no berries, the negative spike in dopamine now occurs when you see a canyon with tall trees. The negative spike in dopamine correlates to the expectation of a bad outcome. You stop the behavior of looking for berry bushes.

You have modified your behavior based on the expectations of the outcomes of that behavior. This is the role that dopamine plays in the shaping of output based on positive and negative results. Behavioral choices are made based on the expectation of positive or negative results and the effect of dopamine within the basal ganglia is the implementing agent of this look ahead neural system.

The most infamous neurological disorder involving the basal ganglia is Parkinson's disease. This disease is caused by the atrophy of the portion of the substantia negra that is the main provider of dopamine. The neurotransmitter dopamine is necessary for normal basal ganglia operation and for learning of new output capabilities, like finding berries. Parkinson' disease causes jerky movements and tremors, the individual parts of an output routine can be exercised but the parts are not strung together properly. The inability to start and stop movements in a coordinated way is also typically a symptom of Parkinson's disease as well as the inability to effect smooth speech. Some alleviation of symptoms can be accomplished by the administration of dopamine precursors that improve the efficiency of dopamine production.

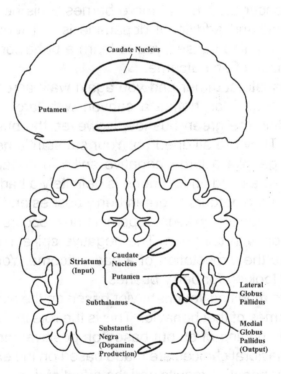

Figure 2-16: Basal Ganglia

Your fourth floor shipping control department contains six rooms that span from the top of the second floor brain stem to just below the fifth floor cerebral cortex. (Ref. Fig. 2-16) The caudate nucleus and putamen together are called the striatum and are the input rooms of the basal ganglia. The next three rooms, the lateral and medial globus pallidus and subthalamus, combine to produce the basal ganglia's output. The last room, the substantia negra, produces the neurotransmitter dopamine that is critical to the operation of the basal ganglia.

The caudate nucleus is an elongated "C" shaped structure that lies over the thalamus with a head in the front and a narrowing tail that sweeps back and down to lie alongside the temporal lobe. The putamen and caudate nucleus are neurologically identical structures. The putamen is separated from the head of the caudate by the massive tract of descending axons from the cerebral cortex to the brain stem and spinal cord.

The putamen and globus pallidus form a lens shaped cone of neurons with the putamen being the outside largest portion adjacent to the insular cortex. The internal portion of this lens shaped structure is made up of the lateral globus pallidus and the medial globus pallidus. The medial globus pallidus is the output room of the basal ganglia and it outputs to the portion of the thalamus that interconnects with the frontal cortex. This medial globus pallidus output controls the reciprocal interconnections between the frontal cortex and the thalamus. The medial and lateral globus pallidus rooms have a fifth room between them, the subthalamus, which lies just below the thalamus and works with them to control basal ganglia output.

The final sixth room substantia negra resides below as part of the midbrain and consists of two portions. The first portion is neurologically similar to the medial globus pallidus output room and projects to the second floor midbrain to control eye movements. The second portion contains masses of darkly colored neurons that utilize dopamine as their axonal neurotransmitter and project to the striatum and frontal cortex.

The first two rooms, the caudate nucleus and putamen, comprise the input portion of the basal ganglia and are made up of cubicles that do nothing but watch the TV screens that show what is going on up on the fifth floor cerebral cortex. Each cubicle in these rooms watches a number of cubicles on the fifth floor. When a cubicle in these two input rooms recognizes something significant on their TV monitors, they signal to the output control rooms, the lateral and medial globus pallidus.

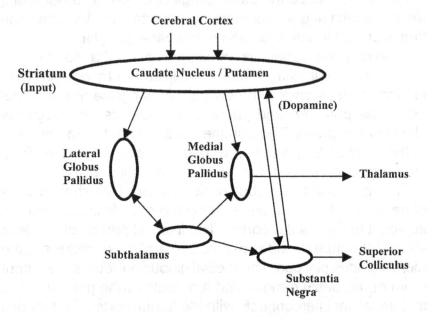

Figure 2-17: Basal Ganglia Nuclei

The striatum is made up almost exclusively of medium spiny neurons. Virtually the entire cerebral cortex, with the exception of sensory input cortex, projects to the striatum in a linear fashion. The striatum is a single layer of medium spiny neurons squashed together in a nucleus that can be thought of as a sheet lying under the cortical sheet, receiving the image of the cortex. These inputs terminate on the dendritic spines of medium spiny neurons. The limbic collar and lower areas of the cortex project to the lower portion of the striatum and the neocortex projects to the upper areas. Cortical input to the putamen is concerned with skeletal motor behavior. The medium spiny neuron in the striatum is the only learning type neuron in the basal ganglia. The neurons that make up the other nuclei of the basal ganglia are hardwired. (Ref. Fig. 2-17)

Medium spiny neurons are the dominant form of neuron within the striatum, both the caudate and putamen. A full 95% of the neurons of the rat striatum are medium spiny neurons. A typical medium spiny neuron contains twenty to sixty separate dendritic branches with approximately 500 input spines per dendritic branch. Thus an average medium spiny neuron contains 10,000 to 30,000 dendritic spines. These spines receive input from the axons of layer five cortical pyramidal neurons. In addition to the cerebral input there are typically from 500 to 5000 dopamine inputs per medium spiny neuron from the substantia negra. Approximately 95% of striatal medium spiny neurons receive dopamine input via synapses on their dendritic spines. These large dendritic arbors are spherical in shape and overlap with one another.

Each striatal cubicle receives cortical input from two cubicle areas of cortex, one area in non-frontal cortex and the area in frontal cortex to which this non-frontal area reciprocally connects. For example, primary motor cortex (M1) and primary somatosensory cortex (S1) concerned with a particular body area both provide input that overlaps in striatal areas concerned with output to that body area.

Striatal medium spiny neurons are normally silent with episodes of firing that typically last for tenths of seconds and can last up to multiple seconds. These neurons are normally silent because it takes a large number of synchronous inputs to cause the neuron to fire. The medium spiny neuron is normally locked in a down state by a continuous shunting current. Synchronous input to a few of the dendritic branches is insufficient to cause the neuron to fire. A large synchronous input that spans most of the dendritic arbor is required to overcome this current. When this condition is met, the shunting current collapses explosively and the neuron enters an up state. The electrical potential of this up state is just below the neuron's firing threshold and the neuron is easily fired by a much smaller synchronous input. This up state is maintained until the input drops significantly to a lower level. The striatal medium spiny neuron is a bi-stable device that exhibits significant hystoresis. In other words, the medium spiny neuron wants to stay in the state it is in. If it is off, it takes a large excitatory input to turn it on; if it is on, a very small amount of excitatory input will keep it on.

The globus pallidus is the main target of both the caudate nucleus and the putamen. This striatal projection is much reduced from the size of the input to the striatum. The globus pallidus nuclei are made up of large neurons that form strongly inhibitory axonal synapses with their target neurons. The medial globus pallidus outputs to the gating nuclei of the thalamus that form thalamocortical loops with the frontal cortex and this strongly inhibitory output inhibits these reciprocal interconnections. The medial globus pallidus strongly inhibitory output causes the reciprocal connection between the thalamus and frontal cerebral cortex to be disabled.

The subthalamus nucleus provides excitatory input to the medial globus pallidus and substantia negra. The substantia negra contains the dark dopamine source neurons that project dopamine producing axons to the striatum and frontal cortex.

Controlling the output of your brain is an extremely important function. How does the basal ganglia accomplish that task? An attempt to answer that question will have to wait for more information to be presented but we can say that the basal ganglia is basically an inhibition structure. You have many behavioral choices available for any situation, stored as patterns in your frontal cortex. Only one behavior pattern is selected and allowed to be operational, all others are inhibited by the basal ganglia.

Each cubicle in the input striatum rooms looks at two TV monitors that show two interconnected areas of the fifth floor cerebral cortex. One area is in the frontal cortex and the other area is in the portion of non-frontal cortex that reciprocally interconnects with that area of frontal cortex. The basal ganglia output controlled by this particular striatum input cube outputs to the portion of the thalamus that reciprocally interconnects with that same area of frontal cortex. That output to the thalamus has the ability to inhibit the reciprocal connection between the thalamus and the frontal cortex area.

If the striatum input cube sees something very interesting on its TV screens, it enters an up state and signals to the basal ganglia output to stop inhibiting the thalamic reciprocal connection to the frontal cortex area. The striatal input cube will continue to remain on and allow the thalamus and frontal cortex area to interconnect until what it sees on its TV screens becomes very boring.

Neural Output – (How You Control Your Skeletal Muscles)

The final subject of our brain building tour, the brain's output system, is a system composed of the neural components we have just visited on each of the five floors. The subject of human neural output includes both skeletal and non-skeletal output and is beyond our scope for this current discussion. We will simplify our focus to a discussion of the output control of skeletal muscles that control movement.

Skeletal movement would seem to be a rather straightforward subject. An extension of the feedback type systems that control blood pressure, body temperature and all the other closed loop systems we have previously covered. If you want to throw a ball to hit a target, it would seem that a series of output commands to the body would be modified by feedback as the action is performed. This is incorrect. Output to skeletal muscles, contrary to common sense, is not a feedback control loop system. The real world is typically to fast for sensory input to be of much help in controlling your skeletal muscles.

An example is catching a ball. If your nervous system relied on sensory input to track and intercept a ball, baseball would be a very high scoring game. What really occurs is that you mentally predict where the ball is going to be in order to move the glove to the right position to intercept the ball. This form of feed forward calculation is commonplace in skeletal muscle output.

Sensory input is also to slow for coordinating complex movements. An example is throwing a ball once it has been caught. Proprioceptive input from the body concerning body part position and muscle contraction is to slow to work as a feedback mechanism. Throwing a ball happens open loop, affected by no feedback. The entire action of throwing the ball is a completely preplanned movement that is fired off and executed ignoring feedback from the body.

The neural program that controls the throwing of a ball is a very large complicated program involving the entire body. This neural program is modified to affect the desired angle and force of release and fired off. The result of the throw is visually received long after the movement is executed. Throwing a ball is not an anomaly. A great deal of voluntary movement is preplanned and executed without feedback. This preplanned voluntary movement is built on top of posture, balance, anti-gravity and other skeletal muscle systems.

The skeletal neural output system is made up of hierarchical layers. Each is a massively parallel controller of skeletal muscle action. Each successive layer builds upon the capabilities of the layer below it. Lower level capabilities are not replaced or circumvented in any way. The spinal cord supports opposing muscle inhibition and provides integration for the skeletal muscle output of the systems above it. The brain stem is the provider of non-voluntary skeletal controls. The cerebellum stores and provides learned skeletal movements. The cerebral cortex, thalamus and basal ganglia form a neural system that determines what voluntary skeletal movement will be executed and drives that movement.

Each level of the output system projects significant amounts of motor output control. These outputs are integrated into a coherent whole by the interneurons of the spinal cord. The planning and selection of a particular voluntary movement is complicated in itself. Because of the hierarchical nature of your output system, the myriad of details that support voluntary movement are not a concern of the cortex.

Your skeletal neural output system is completely linearly organized exactly like your input systems. The primary motor cortex (M1) contains a linear body map and this organization is carried down through the brain stem and spinal cord to individual muscles in a consistent, organized fashion. The somatosensory input and motor output neural systems are interconnected at all levels of the hierarchy.

There are four basic types of skeletal movement. Reflex movements are caused by sudden, extreme levels of sensory input that require an immediate and fast response. These responses are hardwired from input detection to output response. All reflex movement neural circuits are contained in either the spinal cord interneurons or in the nuclei of the brain stem.

Rhythmic movement is what drives breathing and locomotion. The control of these movements is hardwired within nuclei of the brain stem. Rhythmic movement is built on top of the interneurons of the spinal cord that coordinate the excitation and inhibition of opposing sets of muscles. Sensory input is not required for these hard-wired rhythmic circuits to operate. Human locomotion is bipedal and requires a good deal of learned balance control provided by the cerebellum and vestibular nuclei.

The third form of movement is postural and includes elements of balance and antigravity muscle activation. A voluntary movement always requires a concurrent postural movement to counter balance it. The output system in setting up the throwing of a ball must also set up the learned postural counterbalance movement associated with it. The postural movement is actually fired before the voluntary movement. If you lift an object, the muscles involved in the counter balance postural movement will be fired before the destabilizing lift occurs. Again, we see an output system that looks ahead in time to make the appropriate skeletal adjustments in preparation for voluntary movement.

The fourth type of movement is voluntary movement that is driven by the primary motor cortex, M1. After examining the scope of skeletal muscle control in the spinal cord and brain stem required to allow voluntary movement to be smooth and in balance, voluntary movement seems like the simplest part. In a purely mechanical sense, it is. All of the other output systems operate to make voluntary movement possible. This output system exhibits hierarchical control with each successive higher level exerting broader control over the lower level of detail control provided below.

Muscle fiber provides the move in skeletal movement. Each muscle fiber receives some amount of input at all times from various sources. Posture, balance and muscle tone activities are continuously active with voluntary movement superimposed on top of them. The force of a particular muscle cell contraction is directly proportional to the discharge frequency of its driving motor neuron.

Every muscle fiber in your body is innervated by only one motor neuron. Each motor neuron innervates multiple muscle fibers depending on the level of control required for that body area. Motor neurons that move the eyeball innervate four to ten muscle fibers while motor neurons that innervate the buttocks innervate several hundred individual muscle fibers. A motor neuron's effect on the muscle cell is only excitatory and utilizes the neurotransmitter acetylcholine. The collection of motor neurons that innervate an entire skeletal muscle usually project from over one to four spinal cord segments.

The spinal cord and equivalent areas of the brainstem that contain skeletal motor neurons are the first of four levels in the hierarchical output control system. (Ref. Fig. 2-18) Each motor neuron contains approximately 10,000 dendritic synapses that are all reasonably local to the cell body. The large majority of these inputs come from interneurons located in close proximity to the motor neuron. Interneurons link groups of muscles together to facilitate coordinated movement. Input to the motor neuron contains both excitatory and inhibitory input and is quite complicated.

The neural tracts that descend down the spinal cord terminate mainly on interneurons in both the anterior motor and posterior sensory areas. Almost 50% of this input terminates in posterior areas and modulates incoming sensory input. Direct input from the primary motor cortex is the greatest long distance input to the motor neuron. This complex neural spinal cord system of motor neurons and interneurons is entirely hardwired.

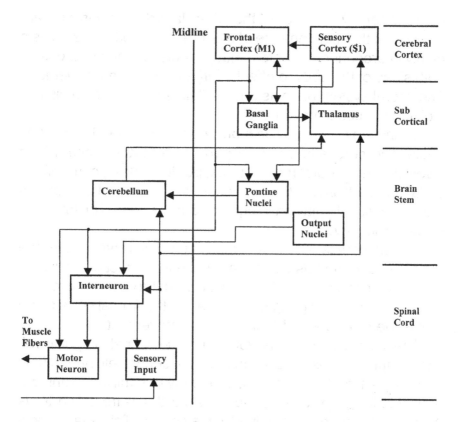

Figure 2-18: Neural Output System

Motor neurons within the spinal cord project bundles of axons that exit each segment of the spinal cord to innervate muscle fibers. The medial portions of these bundles innervate the proximal muscles of the trunk, shoulder and pelvis and the lateral portions innervate the distal muscles of the arms and legs. The medial and lateral portions of these bundles are actually the outputs of two distinct output systems. The older, medial portion is heavily utilized for posture, antigravity, and the other output systems driven from brain stem nuclei. The newer, lateral portion is heavily driven by the cerebral cortex and is the main output component of voluntary movement.

Proprioceptive secondary neurons in the spinal cord provide output to interneurons up to several spinal segments both up and down from their location. Medial secondary sensory neurons tend to project further than lateral neurons and are involved in the maintenance of posture. The medial system is a highly interconnected system that projects to motor neurons on both sides of the body to effect a large coordinated movement. The lateral system projects only ipsilaterally to facilitate distinct individual muscle movement.

Brainstem nuclei provide many varied functions. The pontine nuclei and cerebellum form a part of the output control loop involving the cerebral motor cortex, pontine nuclei, cerebellum, thalamus, and back to the motor cortex. This neural loop allows the cerebellum to play an integral role in the control of voluntary skeletal movement.

The subcortical area is your third output level and contains the thalamus and the basal ganglia. These two structures support the frontal cortex in its role of driving voluntary movement. The thalamus and basal ganglia are a part of the frontal cortex loop that is the main controller of voluntary movement.

The cerebral cortex is your fourth and final output level. Pyramidal neurons in layer five of M1 send their projections directly to motor neurons. Approximately one million axons are projected from the cortex to the spinal cord. The frontal cortex is the newest component involved in the control of skeletal muscle and exerts overall control of voluntary movement. The frontal cortex consists of the primary motor cortex, M1, the premotor cortex (M2) and prefrontal cortex. The axons of large pyramidal neurons in M1, called Betz cells, make up approximately 5% of the corticospinal projection. Axonal collaterals of these Betz cells inhibit the antigravity tone of muscles just prior to a voluntary movement. The motor areas M1 and M2 of the frontal cortex receive a mirror image body projection from the somatosensory areas S1 and S2 of the parietal lobe.

Much of voluntary control of movement is open loop and does not rely on feedback. This implies that motor programs of some complexity must be stored within the brain to enable these output programs to be executed. Voluntary movements are driven from learned patterns and are definitely not hardwired. The brain stores patterns in memory concerning learned output programs. These are called procedural memories that drive behavior and they are stored in the frontal cortex.

Conclusion

Our initial tour of your brain building is now complete. We have visited each of the five floors and taken a brief look at the various rooms and their functions. That brief look has covered a good deal of information. You now know what components comprise your brain, where they are located, some of their neural wiring, how sensory input is received and processed, and how your brain controls skeletal output. Next we will utilize your new knowledge to explore some of the functions provided by your brain.

Chapter 3

Lets Think About What The Brain Does

Introduction

In our brain building tour, all of the various components that comprise your brain were discussed. Next, we will examine some of the higher-level functions and neural characteristics enabled by those components. All brains provide a variety of capabilities for the animals they serve. The capabilities supported by the human brain dwarfs the mental capacity achieved by all other animals. The term dwarfs does not do justice to the disparity between human intellect and all other living things on earth.

Your brain contains approximately one square foot of cortex. Your two dimensional cortical sheet is both wider and longer than all other animals. Being wider allows you to recognize more objects. Being longer allows you to support more cascaded layers of pattern detection. Remember those fifth floor cortical cubicles that pattern detect their input and pass the results on to other areas of cubicles. Those cascaded levels of pattern detection allow the building of relationships between objects. They enable intelligence.

Memory is the storage and recognition of patterns within the cerebral cortex. The recognition of memory patterns in the human brain supports two distinct forms of learning, behavioral and knowledge. The level of behavioral complexity demonstrated by humans is enabled by the large amount of frontal cortex dedicated to storing behavioral patterns. The many layers of cascaded pattern detection supported by non-frontal cortex enable the complexity and variety of knowledge accessible to humans. Humans recognize a vast array of objects and many complex relationships concerning those objects. Language, a distinctly human endeavor, clearly illustrates the ability of the human brain to build and recognize abstractions and complex associations between objects. The cascaded levels of pattern detection supported by the human brain enable this uniquely human capability.

The last subject of this chapter contains a highly speculative examination of the evolution of the human brain. The evolution of the human nervous system has been an additive process. Operational components have not been replaced as new neural structures have been added on top of them. The evidence for this additive-layered approach exists within the neural systems of the various animals that have survived intact to this day with their various neural designs. This existing continuum of primitive to more advanced animals provides phylogenetic evidence that allows us to examine the neural capabilities of each stage of neural system design. Your body contains each of these older neural systems, like the pages of a neural evolution book. Assuming that the evolutionary pages contained in your neural system are an accurate reflection of the evolution of the human brain, as I do in this discussion, must be viewed with skepticism.

This circumstantial evidence does not allow this discussion of brain evolution to be anything but 100% speculative. That statement leads to an obvious question, why do it? This examination of the growth of the human brain to its current level of capability allows us to reflect on the new functionality that each new component provided and hopefully gives us insight as to its present function.

Memory

Memory is the ability to store and retrieve information about what you have experienced and what you have learned. Memory is the mechanism that allows you to retain knowledge. There are four distinct classifications of memory, short-term memory, long-term memory, declarative memory and procedural memory. Short-term memory is the ability to hold and retain a number of things in your head long enough to accomplish a short-term task. Receiving a phone number from the information operator and remembering it long enough to dial it is the classic short-term memory example. Long-term memory is what we can recall from the past. This is permanent memory storage that degrades with time and non-use. Declarative memory describes the memories you have for objects and space and procedural memory refers to storing learned behavior.

Your short-term memory ability to hold a small amount of unlearned information in your conscious awareness is usually for the purpose of accomplishing an immediate goal. The information is typically discarded and not stored as a long-term memory. There is a limit to the amount of information you can hold in this manner. The limit seems to be seven things plus or minus two, depending on the individual. You can usually remember a seven digit telephone number long enough to dial it once. If you miss dial, the number may have to be re-accessed. The items you can retain via short-term memory are not limited to simple numbers or things. They can be complex collections of previously stored information or other similar bound items. The limit is still around seven, seven numbers or seven poems or seven countries.

Long-term memory is permanent storage. There is a gradual loss of recall over time depending on usage. A memory is recognized when new input patterns compare with a previously stored pattern. The stored memory pattern is refreshed and at the same time altered to incorporate the new input that caused it to be recognized. Frequently recalled memories are more easily recalled with usage.

The storage of people, places, and things is termed declarative memory. Lets combine people and things into objects and substitute space for places in our discussion of declarative memory. This is convenient because the cerebral storage locations for objects and space are distinct and different. The temporal lobe on the lower side of the head is the storage location for objects. The parietal lobe in the upper back of the head is the storage location for spatial perception.

Declarative memory storage space is all non-frontal association cerebral cortex. What you store as declarative memories are patterns representing space, objects and associations between objects. Humans store an amazing variety of information in declarative memory. History, geography, language, and scores of other learned information. Your entire life's knowledge has been stored within your non-frontal cerebral cortex. Much of this information is lost over time with disuse. This is a tremendous amount of information to be stored in a storage facility so small.

Memory is implemented by neurons and neurons only perform pattern detection. It is therefore not surprising that memory tasks that involve simple pattern detection allow your memory system to operate at peak efficiency. Remember the visual recall task described in the visual input section. A series of slides containing pictures were shown once to people for a few seconds at intervals of approximately five seconds. Two days later, the subjects were shown a series of two slides, one new and one from the list of previously viewed slides. The subject indicated which slide seemed more familiar. The error rates for all subjects were extremely low even when the number of slides shown numbered 10,000. One brief visual encounter allowed enough pattern recognition to enter long term memory to make this an easy memory task. This argues strongly that you store everything at some level within your memory.

Procedural memories store output behavior. A typical skeletal procedural memory is "how to ride a bike". Riding a bike is a very complex task. Learning to control steering, apply proper pedal pressure, braking and most importantly, maintaining balance requires a good deal of practice. Learning to ride requires surviving a large number of frightening failures. Mastering riding a bike takes effort, motivation and most importantly, persistence. Building procedural memories is a slow additive process.

If you want to learn a new cord on the piano, it will take around 150 tries to basically get it right. The first try will probably be a disaster followed by slow improvement with each successive try. Declarative memory formation is incredibly fast. 10,000 pictures remembered with one viewing is as fast as it gets. One trial ride on a bicycle only begins the process of building the procedural memory that will enable you to ride a bike. Once learned, you will remember how to ride a bike for the rest of your life. Procedural memories are stored in frontal cortex and drive human behavior. Not just skeletal movement but all of human behavior.

Some clues about memory storage come from events that disrupt it. A blow to the head that renders a person unconscious always removes recollection of the few seconds prior to the blow. A fighter never remembers if it was a right or a left that knocked him out. A typical treatment for psychosis used to be electroconvulsive shock treatments. These treatments always caused a retrograde amnesia of approximately 30 minutes. Apparently at least two memory storage processes exist. The first requires a few seconds for memory consolidation and the second approximately a half hour. After a few hours, long-term consolidation of memory seems to take place that makes memory somewhat permanent.

The synthesis of protein is required for the storage of memories. If protein manufacture within the neuron cells of animals is inhibited, memory formation does not take place and learning is disabled. Proteins are the basic building blocks of cells; this implies that structural changes in neurons are required for the storage of long-term memories.

Attention and emotional state are the gateways of memory storage. The attention paid to incoming information and the emotional state of the observer are both primary factors in determining the ability to recall the information at a later date. If you are not paying attention to a particular sensory input stimulus, you will not be able to recall the stimulus. The strength with which a memory is stored and accessible for recall is directly proportional to the emotional state at the time of the event. Extreme emotional states during events cause particularly strong memories of those events to be formed that are easily accessed or remembered.

A memory is enabled because pyramidal neurons in the cerebral cortex recognize a previously encountered pattern and fire to signal that recognition. The new pattern that causes the recognition becomes a part of that memory pattern. The recalling of a memory causes your perception of that memory to be altered due to the inclusion of the new input. Memory is a process where new input causes you to retrieve previously stored older input and older stored input is utilized to measure and interpret the new input.

Learning

Learning is typically defined as behavioral change based on experience or the process of acquiring knowledge. This definition reflects the fact that there are two different kinds of learning, behavioral and knowledge. Behavioral change involves the creation of procedural memories stored in the frontal cortex; acquiring knowledge involves the creation of declarative memories stored in non-frontal cortex. Understanding how new declarative and procedural memory patterns are created and modified within your cerebral cortex is the equivalent of understanding how you learn.

Acquiring knowledge involves the building of associations between objects in declarative memory. The visual image of an apple, the taste of an apple, the smell of an apple, the written word apple, and the spoken word apple are all separate objects stored in different areas of your declarative cortex that are somehow linked. Sensory input that causes you to become conscious of any one of these apple objects will also cause the other object representations of apple to be accessed within declarative memory.

Words and numbers are objects stored within your declarative cortex that are abstractions of other objects. These abstraction objects are somehow linked with the declarative pattern representations of the objects they represent. Only humans have enough cerebral cortex to build these complex associations between declarative objects that support language and mathematics.

When you meet a new person and hear their name, the visual pattern of their face and the audio pattern of their name are somehow linked. The visual pattern of their face is stored in the rear temporal lobe and the audio pattern of their name is stored in the upper temporal lobe. Understanding how these associations between objects are built, stored and accessed is required for an understanding of how you acquire knowledge.

Behavioral learning implies new behavior or the creation of new procedural memory patterns. Your frontal cortex stores the sum of your behavioral repertoire. This collection of procedural memories has been stored over your entire lifetime. Human behavior is quite complex and procedural memories are modified and extended in response to each declarative pattern that causes them to be recognized. Both declarative input and limbic emotional input play a role in the recognition and selection of procedural memories.

Intelligence

Intelligence Quotient (IQ) tests have been trying to quantify intelligence for some time. An IQ test measures both a person's memory capability and their reasoning capability. Memory is tested by requiring recall of information that should have been learned and retained through education. This information includes facts, mathematical abstractions and vocabulary. Reasoning capability is measured by presenting some information and asking a question that requires some abstract associations about the information to be worked out in order to provide the correct answer. The test is measuring how well this human's brain builds these unique relationships between the information presented. The test requires applying real world knowledge to a problem and working out the answer. It is the building and recall of neural associations that facilitates both knowledge and reasoning ability.

The amount of gray matter in the cerebral cortex of a brain is directly proportional to the mass of the neurons within the cortex. The amount of gray matter in the cortex of a brain is also a direct indicator as to the intelligence enabled by that brain. This is true in animals and humans. Persons with higher IQs have more gray matter than individuals with lower IQs. What is counter intuitive is that the persons with more gray matter and higher IQs have less neurons. Their cerebral neurons are larger and have significantly larger and more complex dendritic trees. Learning that enables intelligence results in more neuron death and less neurons. The competition between neurons to survive is accentuated by learning.

The higher-order aspects of human intelligence are enabled by the large amounts of association cortex in the human brain. The large association areas of the parietal and temporal lobes are directly connected with the matching large area of frontal association cortex. These large association areas represent the largest cortical differentiation between humans and other mammals, including primates. These large interconnected areas of declarative and procedural association cortex are what allow you to read this book.

Language / Math

Language and math represent a form of learning that is both very structural and additive. Additive in the sense that simple associative building blocks are successively built up into ever more complex constructs. The basic building blocks and construction rules for language and math must be learned and stored in declarative memory prior to puberty or these skills will never be learned. Children never exposed to language prior to adulthood will never master language and the same is true for math. Language and math require a great number of neural objects to be stored and complex associations between those objects to also be stored. The young human brain contains a vast amount of neurons and synapses. The adult human brain does not contain enough white space in its cerebral cortex to allow these subjects to be mastered from the start.

Very specific regions of your cerebral cortex enable language. Much of this knowledge has been derived by the observation of people with damage to these regions of cortex and noting their language disabilities. The left temporal lobe and regions of the frontal and parietal cortices are critical in enabling human language. The left temporal lobe is where you store the neural representations of word objects. Damage to the left temporal lobe usually has a deleterious effect on language ability.

Wernicke's area refers to a particular lower portion of the parietal lobe that is adjacent to the auditory cortex and is instrumental in word recognition. Broca's area is a region of the frontal cortex forward of the motor area controlling the mouth and tongue. Broca's area is responsible for speech and the construction of sentences. A stroke that damages Broca's area can severely disrupt the ability to form correct sentences and may cause a patient to only have the capability to utter nouns for instance. Stroke damage to Wernicke's area leaves a patient's sentence structure in place but may cause the patient to select the wrong words.

Human language capability includes speaking, hearing, writing and reading. Each of these activities represents different input and output manifestations of the same basic language capability. Words are abstractions of objects, actions and attributes. In order for a word to serve as an abstraction for another thing it must be associated neurally with that thing in the brain. Language is unique to humans and many argue that language enables all of the higher order functions that we classify as human behavior. The human brain is the only brain that contains the amount of association cortex required to build the associations necessary to support language and mathematics. Many humans have the ability to communicate in more than one language. The cerebral storage location for the same word in different languages occupies different areas of temporal cortex. The grammatical framework for different languages is stored in different areas of Broca's area.

Consciousness

The subject of consciousness has been debated for thousands of years. Consciousness is a very ambiguous term and ambiguity typically leads to disagreement. The problem with consciousness is that we all subjectively know what it is because we are conscious ourselves but it is not directly measurable or observable. There have been numerous attempts to quantify and define what is meant when we say something is conscious; awareness, arousal, alertness, and attention are often used as defining terms. Our famous epileptic patient HM was completely conscious even though he could not form new memories. Apparently memory formation is not a prerequisite for consciousness.

You are not conscious of information manipulation within your primary sensory cortices. Your primary visual cortex, V1, breaks visual input into component lines and colors for each area of the retina and provides that information to secondary visual cortex for image construction. You are only conscious of the reconstructed visual image of the space and objects within your view. These higher-level visual images are constructed piecemeal through secondary cortex to final construction in association cortex. Primary sensory input areas do not project to prefrontal cortex, the hippocampus or to the basal ganglia. Secondary sensory input cortex projects sparsely to these areas and association cortex projects profusely to these neural structures.

In order for consciousness to exist within the brain, a certain number of neural components must be operational. These are basically reticular type areas that control your level of alertness. The projections from the intralaminar nucleus of the thalamus must be intact in order for consciousness to exist. The reticular formation of the brain stem must also be operational or the brain lapses into coma. These reticular systems are not on and off switches. They provide a continuum of stimulation to specific neural areas. These reticular systems are definitely required for awareness that is a prerequisite of consciousness.

Right vs. Left Brain

There is differentiation in functional responsibility between the left and right brain of humans. The left brain controls the right side of the body, controls language, and has been described as the intellectual, rational, verbal, analytical brain. The right brain controls the left side of the body, deals with the emotional aspects of communication, and has been described as the emotional, nonverbal, intuitive brain. The left-brain is called dominant and most humans are right handed. Left-handed people are less likely to have their language center on the left side of their brain.

Some patients with severe epileptic seizures have had an operation called a split-brain procedure. In this surgery, the entire corpus callosum and all other axonal interconnections between the left and right brains are severed. This leaves a person with two disconnected cerebral hemispheres and thus two separate brains. Testing of problems presented to each side separately verifies that the right brain is mute. These patients cannot write with their left hand. All of the various attributes attributed to each of the two hemispheres have been verified in these patients. Even with all the evidence for differentiation between the brain halves, in a normal brain the halves are extensively interconnected and operate as a whole.

Sleep

Sleep is a behavior that exists in most animals, even fruit flies sleep. Why animals sleep is still a mystery but it must satisfy a very old and common need. The sleep patterns of all humans are consistent and involve four stages of sleep. The determination of these four stages is the EEG pattern or brain waves of sleeping persons. Stage one sleep is characterized by low voltage, high frequency, and somewhat disorganized wave patterns. Each successive stage of sleep, stages two to four, are characterized by slower frequency, more organized waves and higher voltage. Stage four is recognized by consistent high voltage brain waves at approximately 40 cycles per second. Stage one is light sleep and stage four is very deep sleep. Human sleep starts at stage one and cycles to stage four over an approximate time of 45 minutes. The cycle then continues with a reverse transition to stage one sleep over another 45 minutes. Thus a complete cycle from one to four and back is around 90 to 100 minutes. A typical nights sleep will involve four to five complete cycles or from six to just over eight hours of sleep.

The aspect of sleep that has received the most research is dreaming. Dreaming occurs after you cycle back up to enter stage one sleep. During stage one sleep, dreaming episodes are characterized by rapid eye movements (REM) and are called REM sleep. The EEG during stage one sleep is closest to the EEG of awake individuals. During dreaming episodes all skeletal muscles are inhibited by descending projections of the brainstem reticular formation with the exception of eye muscles, middle ear muscles and respiration muscles. The lack of inhibition to the eyes enables the rapid eye movements during dreaming that are characterized in the name.

During REM sleep, the neocortex exhibits the desynched waveform of stage one sleep but the hippocampal EEG remains highly synched at a very slow four to ten cycles per second. As you sleep through the four or five up and down cycles through the night, the stage one-period and REM episodes get longer with each cycle. Everyone dreams and exhibits REM sleep unless they are on medication that inhibits dreaming. Patients on medication for years that inhibits dreaming have experienced no harmful side effects. Premature human babies exhibit REM dreaming for up to 80% of their sleep total. Full term babies exhibit 50% REM sleep out of their normal 16 hours of sleep per day. REM sleep continues to decline to 25% at age ten and then remains at that level until old age. Stage four deep sleep also declines with age and can completely disappear by age 60.

Dreaming is an interesting state. The cerebral cortex consumes more oxygen during dreaming than in the alert, waking state. Dreaming is mostly visual with some auditory and somatosensory sensations and little or no taste and smell sensations. People who become blind lose the visual aspects of their dreams over time. There is some evidence that learning is somewhat inhibited in humans if REM sleep is continually disrupted.

Sleep occupies one third of your circadian rhythm controlled by the hypothalamus. The actual neurotransmitters that control arousal and therefore sleep are norepinephrine and serotonin that are delivered throughout the brain by nuclei within the brainstem.

Human Brain Evolution

We are now going to spend some time exploring the evolution of the human brain. This is pure speculation as there is absolutely no surviving physical evidence pertaining to the evolution of the human brain. We do have the phylogenetic evidence of the neural systems of all the various animals existing today. The neural systems of these animals have not stopped evolving in order to provide us an audit trail as to your brain's evolution but there is a consistent layered structure of the same neural components contained within the neural systems of all animals.

Animals we humans consider "lower" than ourselves only have the neural components consistent with their position within the phylogenetic tree. The neural components an animal possesses each performs an analogous function to the function it performs in the human neural system. The neurons that make up the various components and the neural wiring of those components are consistent across species. Even the neurotransmitters utilized are typically the same. This gives us some, not total, assurance that we can extrapolate from simpler neural systems to our own. It is also indicative that the evolution of brains is recorded in the neural systems of animals that comprise the phylogenetic tree.

In the evolutionary growth of animals, entire sequences of groups of genes are sometimes replicated and cause an extra copy of the body part specified to exist in the animal. These extra body parts are then differentiated over time to improve the survivability of the animal. The genetic control of the segmented bodies of fruit flies is strong evidence for this type of evolutionary growth. We are going to postulate that this happens repeatedly in the evolution of the brain. Replication of genetic sequences that add neural material on top of the existing nervous system will be assumed. When we have completed examining one stage of the growth of the nervous system, we will add neurons for the next layer and examine their differentiation.

Evolutionary change results in new species and also generates substantial change within a species. The body plan of a species can be altered quite quickly given enough environmental pressure. Domesticated dog varieties are an obvious example enabled by human pressure. Some extreme examples of evolutionary body change driven by environmental pressure are the trunks of elephants, the necks of giraffes, and the cerebral cortices of humans.

There is one other aspect of brain evolution that enables us to make educated guesses about how the brain evolved. Brain evolution is strictly an additive process. Older neural structures remain in place as new structures are added. The function of the older structure continues to operate as the new structure differentiates. Each evolutionary stage represents a complete functioning nervous system. The human brain was not designed, it was grown.

One event in the evolution of life both enabled and required the kind of control that neurons provide. That was the appearance of multicellular life over 500 million years ago. Genetic specification of cellular differentiation within an individual life form was a large leap in genetic capability and led to an explosion of new life forms. The Coelenterate phylum includes the Hydra and is the most primitive phylum to include a rudimentary nervous system. This phylum has existed since the dawn of multicelled animals on earth. Life with a brainstem existed over 500 million years ago. Life with limbic systems first appeared between 200 and 300 million years ago. The cerebral cortex first appeared between 100 and 200 million years ago. Brain evolution has been going on for some time.

Genetic coded cellular differentiation allowed shells, cartilage, and muscle cells to evolve. Muscle cells multiplied and formed a pump to drive fluid among the cells. Other muscle cells formed both in parallel and in series to enable motion.

We will assume that the first neuron evolved from a genetic change that caused a differentiated muscle cell. Perhaps a sensitivity to light detected by a muscle cell allowed the animal to swim to the light and increase its survival odds. This first neural ancestral cell was a combination of sensory input and muscle cell. This cell differentiated into a neuron signaling to a muscle cell. That first neuron combined the capability of a sensory neuron and a motor neuron. Some control was good and more control was better in increasing survival odds and neurons evolved rapidly. These newly evolved rudimentary neurons were arranged as a one-cell layer of control detecting input and signaling directly to muscle cells.

Once genetically differentiated from muscle, the neuron was free to evolve and expand in many directions in order to improve the control of muscles in the survival game. Through all of subsequent expansion and evolution, nervous system output has continued to be tied to muscle cells. The nervous system has improved its control but remains slave to its original intent. Neuron genes were replicated and groups of parallel neurons were formed. At some point the whole single layer of neurons was replicated causing two neurons to be in series. This allowed input and output differentiation to begin.

Neuron to neuron communication was then required as neurons were now signaling to other neurons and not just to muscle cells. The signaling neurotransmitter mechanism between neurons and muscle cells, acetylcholine, is still the most widely utilized inter neuron signaling mechanism in the entire human brain. Through the same process, now add another layer of neurons and we have our first interneurons. The addition of a third layer of neurons allowed neuron differentiation not tied to input or output. This was the start of the human brain. These first interneurons allowed more complicated control of groups of muscles supporting more complex body types.

The growth and evolution of input neurons, interneurons and output neurons diverged. Input neurons differentiated into the wide variety of sensory neuron types we see today. We have multiple sensory modalities and a variety of sensory neuron types within each modality. Output motor neurons have not changed in function. Each muscle cell is still innervated by only one output neuron. The number of output neurons has grown in proportion to the amount of muscle fibers that require innervating. The greatest change with evolution has occurred in interneurons that have evolved into the human brain.

We now have three layers of neurons. The greatest opportunity for improved probability of survival and procreation via the control of muscles lies with the interneurons. Speculation about the evolution of interneurons is what concerns us now. Let a small amount of time elapse, perhaps 20 or 30 million years. Interneurons have come a long way. In the ancestors of vertebrates, chordates, they fill the spinal cord with sophisticated input and output control. Through genetic change they have grown faster than the bony spinal cord that housed them and have formed a rudimentary brain end. Differentiated sensory input other than body related input is concentrated near this clump of interneurons.

This simple brain like collection of interneurons provides output functions relevant to more complicated body types. These output controls project to the interneurons in the spinal cord and enhance but do not alter their capabilities. The body's sensory input paths extend into this new collection of interneurons. Modulation of sensory input and utilization of sensory input for output control are provided. Every successive neural layer of interneurons that will be added through evolution will provide interconnection of sensory input and motor output. The motor output paths and sensory modulation paths will be extended with each additional neural component. Interneurons will continue to serve their original functions of sensory modulation and motor output, even as they evolve into the human brain.

At this stage of evolution, animals are hardwired for one behavior and one behavior only. They are the ultimate one trick pony. The earth at this time is a behavioral white space. Each new neural genetic change creates another slightly different behaving animal and almost everything works. There is no competition for most new behaviors and genetic exploration and colonization of the behavioral white space occurs geologically quickly. Since we are in complete speculation mode, lets assume that this creates the Cambrian explosion of life forms. Viable behavioral niches are colonized and body types are modified more slowly to defend the space.

Neural material continues to be added to the top of the brainstem. We now have a variety of animals with rudimentary brainstems. Those brainstems have evolved to perform all of the bodily housekeeping functions that are the duty of brainstems today and implement one hardwired behavior. This level of brain evolution held sway for many millions of years and a large number of very complicated hardwired behaviors evolved in that long period.

One hardwired behavior is extremely restrictive. There are lots of useful types of behaviors required for different situations. It would be very beneficial if the animal could switch between hardwired behaviors. Finding food, finding a mate, and raising young are just a few of the hardwired behaviors it would be useful to switch between.

The next innovation of the evolving brain solved this problem and the limbic system was the answer. The solution involved controlling the switching between hardwired behaviors through the use of chemicals. The control of the chemical or hormonal state of the body evolved to allow the animal to express different hardwired behaviors. Whole new layers of hardwired neurons were added on top of the brain stem to accomplish this feat.

Watch a reptile build a nest, hunt, or feed its young to witness this type of neural control. This is a large step forward in the evolution of the brain. Quite complex behaviors can now be genetically explored and only switched on when required. Life now has subroutine capability. This is a large innovation and holds sway for perhaps 100 to 200 million years and is still a quite successful neural model for many animals that exist today.

The neural system at this point allows animals to grow quite large and exhibit very complex behaviors. The problem is that genetic change of behavior is very slow in real time. Clipping off 100 million year chunks makes it seem fast but it is really glacial. In hindsight, we know what is required. These animals need to have the ability to learn. Having the ability to learn is going to require a completely new type of neural structure. A neural structure whose neurons react to incoming patterns and are essentially rewired as a result of firing in response to those patterns. A genetically specified regular structure that will become wired with use. How layers of neurons were replicated and then differentiated to become this new type of neural structure is completely unknown. We do not know how it happened but we do know that it happened.

The first neural structure to appear that was dedicated to learning was the cerebellum. The problem was simple. It is not possible to completely hardwire all of the balance requirements that movement creates. The brainstem vestibular nuclei had already evolved to provide hardwired output to motor neurons to control balance but it was not a sufficient solution. Movement in general allows animals to enter new environments and greater freedom of motion allows exploration of those environments. Genetic variation is not going to solve this problem. An animal needs to be able to learn to balance movements. A deer is prepared to learn to walk at birth and a human at around a year old but they both need to learn.

The neural structure that enabled this learning started out as additional layers of neurons on top of the vestibular nuclei that formed a simple three-neuron circuit. That circuit interacted with the hardwired vestibular neurons to allow the balance associated with a particular movement to be learned. The effect was very small at first but very beneficial. The neural layer differentiated into a three-neuron sheet of identical neural circuits that exists today as the archicerebellar lobe of the cerebellum.

This simple problem led to a very elegant neural solution. Neurons within the fledgling cerebellum would fire when they recognized body and balance input patterns. The strength of their firing was proportional to the closeness of the input pattern comparison to the learned pattern. Their output caused the vestibular nuclei to modify its motor balance output. This learning of patterns within the new structured cerebellum allowed balance for movement to be modified based on learning. This simple beginning will be replicated to build all of the learning structures that exist within the modern human brain.

The cerebellum expanded in surface area in conjunction with the evolution of brainstem motor output nuclei. This expansion formed the paleocerebellar area that provided learned feedback to the brainstem motor output nuclei based on proprioceptive input.

The cerebellum then differentiated to form a new learning structure that solved a different learning problem. The problem was how to modify limbic controlled behavior based on learning and the solution became the hippocampus and limbic cortex. Limbic control had evolved to a high level of sophistication but was still completely hardwired. The solution the early hippocampus and limbic cortex provided was to store object and behavioral patterns in cortical memory and associate those patterns with the limbic-based emotional state that existed when they were stored.

Lets take a look at an example. An animal with limited new cortex and hippocampus encounters a predator while drinking from a stream. The predator's attack is survived by flight and causes a great deal of limbic generated fear. The pattern of the predator and the behavior of drinking from a stream are stored in cortex. When those patterns are recognized, the hippocampus causes the limbic system to project the emotion of fear. The hippocampus causes the limbic system to generate the emotional state that existed when those memory patterns were stored. Seeing another like predator or drinking from a stream generates fear that causes the animal to behave in a very cautious manner.

The modern hippocampus still has the three-layered structure closely resembling the cerebellum and continues to provide the same limbic learning control function. These three layer structures have main learning pattern neurons within their middle layer supported by hardwired neuron layers. We now have evolved up to an animal that can associate limbic controlled behavior with learned object and behavior patterns. We are up to an animal such as a crocodile on the tree of life.

The cerebellum and hippocampus divide into separate components and the stage is set for the next step forward. Neural evolution up to this stage has been through the addition of vertical layers of neurons followed by differentiation of those neurons into different nuclei. With regular learning type structures the growth is mainly horizontal. These structures expand into sheets of regular neural circuits and their inputs, projections, and supporting neural structures expand in step. The cerebellum has expanded horizontally through time but it has remained a collection of identical separate parallel neural circuits.

The hippocampus and early limbic cortex followed a slightly different evolutionary path as they expanded horizontally. The three layers of the hippocampus differentiated into a more complex six-layered structure in the newer areas. Existing areas of cerebral cortex always established reciprocal connections with newer areas as the cortical sheet expanded. This reciprocal communication allowed newer cortex areas to form learned patterns from the learned patterns of older areas. The growth of what was essentially limbic association cortex was paralleled by the growth of another neural component, the thalamus. The gating control thalamic circuits appeared with the first appearance of non-hippocampal cortex, the cingulate gyrus, and have continued to grow to match the growth of the cerebral cortex. We now have animals that can learn quite complex patterns of behavioral limbic control.

The primitive cortex continues to expand and neocortex begins to appear. As the cortex expands, thalamic gating neurons expand to interconnect with the new area, and reciprocal connections between new and older cortex are a part of the expansion. One consequence of these reciprocal connections between old and newer cortical areas is that the hippocampus maintains communication through the parahippocampal cortex with all new areas as they are formed. The first function performed by the small new areas of cortex is primary sensory pattern detection. One of these early primary sensory cortex areas is the somatosensory cortex S1. The modulation of sensory input now moves up to S1 as reciprocal projections are targeted to the interneurons of the spinal cord.

S1 cortex now expands and begins to differentiate into what will become the primary motor cortex M1. The projections of S1 continue to modulate spinal input but also now begin to drive motor neurons. In humans, this duel projection of sensory modulation and motor control divides into two distinct areas. The older S1 cortex actually becomes M1 and newer cortex takes over the role of S1.

All of the primary sensory cortices now make their appearance. The brainstem nuclei that support auditory and visual control remain in place and interconnect with the new cortex through the thalamus. Secondary sensory cortex and then association cortex that performs pattern detection of multiple sensory inputs appear. The expanding cortex now continues to provide more association cortex that enables cascaded pattern detection. The basic cerebral circuit remains the same throughout this and future expansion. Thalamic gating neurons form reciprocal connections with each new area and the hippocampus maintains communication through the parahippocampal cortex.

A new layer of neurons that will become the basal ganglia makes its appearance. The ability to learn motor patterns allows a great expansion in the possible number of motor patterns. A neural mechanism must exist that unifies and serializes the neural patterns driving skeletal muscles. This neural function received input from the cortex and controlled procedural memory through the thalamic connections with the frontal cortex. The basal ganglia expand and differentiate to interconnect with the expanding declarative and frontal cortex. The globus pallidus output expands to match the growth of the thalamic connections with the growing frontal cortex. The input portion of the basal ganglia differentiates into the striatum and grows to match the growth of association cortex. We are now up to the level of an opossum on the evolutionary tree.

The basic design of the brain is now in place. All that remains is expansion that enables ever-greater forms of learning. Cortex grows larger association areas enabling more complex recognition of input. The basic cortical neural circuit remains the same and the thalamus grows and interconnects to keep pace. The motor cortex expands to premotor and then to association prefrontal cortex. The thalamus and basal ganglia grow and interconnect with the new cortex. The cerebellum expands greatly at this point with the addition of the neocerebellar lobe that provides input to M1 through the thalamus.

With expansion of the cerebral cortical sheet, the thalamus divides into separate nuclei serving each area. The hippocampus supports reciprocal interconnection with the new areas through the parahippocampal cortex. The basal ganglia striatum expands to support input from all of the newer cortical areas. The modern brain architecture is now complete. It has been complete for a very long time.

All mammals have the full compliment of hippocampus, limbic cortex, primary and secondary cortices for visual input, primary auditory cortex, primary somatosensory cortex, gustatory cortex, primary motor cortex, premotor cortex, and some temporal cortex devoted to object recognition. Mammals have been around for about 100 million years. From the earliest mammals, evolution has taken 99 plus million years to produce the human brain. The human brain is basically a very large extension of cortical area based on this same basic design.

It is interesting to speculate about what caused the human brain to greatly expand its cortical sheet. At some point in primate evolution the amount of association cortex enabled a very crude form of language. Language research with chimpanzees has clearly shown the rudimentary capability to associate both verbal sounds and written symbols with objects. At some point in human evolution the ability to communicate verbally became essential to the survival of the individual. Once you can associate a sound with an object, the only things inhibiting having sounds for all objects is the amount of association cortex available to learn the patterns required and the ability to vocalize the different number of sounds required. Language ability is what clearly differentiates humans from other mammals and language may be the evolutionary driving force that lead to the expansion of the human cortex.

This discussion of brain evolution is complete speculation. This speculative view of the capability provided by each additional neural component is hopefully useful in understanding the modern function of neural components. Existing components have remained in place and operational during the complete 500 plus million years of brain evolution. New component function modulates older capabilities but it does not replace them. This has created the audit trail existing within the phylogenetic tree that was used in this speculative discussion.

Conclusion

This concludes our first pass in the exploration of your nervous system. Memory, intelligence, learning, and consciousness are all high-level neural capabilities that define what it means to be human. In the following chapters we will explore each of these subjects again starting with the knowledge recently gained. The goal of the following chapters is to explore what we know in order to apply analysis as to the functionality of neural components. Finally, we will endeavor to explain the higher-level human capabilities explored in this chapter.

Chapter 4

If Neurons Are The Solution, What's The Problem?

Introduction

Remember our neuron brick, the active component of all of the neural structures and circuits we have been describing. Each individual neuron is a pattern recognition component that is isolated from the big picture. Even with a prodigious number of inputs, each neuron is a very small piece of any one brain component. The death of any one particular neuron will have no measurable effect on the performance of the whole. The brain is a massively parallel device with inherent redundancy.

Lets imagine what it might be like to be a pyramidal neuron in the cerebral cortex of a human brain. I am a pyramidal neuron. I wake up every day to find myself in a stadium with 60,000 people in it. These people stand up and then sit down. Some jump up and down many times a second. If a person never stands up or I don't see that person as part of a pattern that I recognize, after a while, a trap door opens underneath their seat and they are gone. If one section of the stadium is especially active and I recognize a pattern to what they are doing, more seats can be created to allow more people to represent what this group of people seems to be excited about, whatever that is. Adding more people like this helps me recognize when they are excited.

My job is to watch these people jump up and down and push a button if I see a pattern that I have seen before. If I think I recognize a pattern, I push my button. If I know I recognize a pattern, I push like crazy. Every time I push my button, 20,000 people jump up in some other stadiums just like mine. I do not know, or care, if the 60,000 people in my stadium are controlled by 60,000 different stadiums or less, they all look the same to me. I also do not care where the 20,000 people are that jump up when I push my button.

I push for as many patterns as I recognize. I can learn new patterns. The patterns can involve all 60,000 people but I also react to patterns in just part of the stadium. When I was first born, I did not recognize anything. I signaled when I thought I saw something and luckily the other stadiums used my input. If they had not used my input they would have trap doored all my people, next they would have revoked my season pass.

This fanciful description of what a pyramidal neuron does has a problem. A person with visual input and a whole brain to interpret that input is not a valid analogy of how a neuron works. This description of our neuronal brick does not match the evidence from neuronal research. A neuron fires when its excitatory inputs are greater than its inhibitory inputs by the threshold amount. In order to be true to the evidence, we need to have people shout as they stand and replace the person in the center of the stadium with a noise meter that signals when the noise is above a threshold.

When a pattern of standing people is recognized, those people are given a megaphone. They get to make a little more noise and their probability of being recognized as a group is enhanced. In our stadium analogy, a full memory pattern is recognized over vast numbers of stadiums with each stadium's contribution slightly enhanced for its portion of the total pattern.

The sum total of all those enhancements represents a significant total improvement for recognition of the entire pattern.

In this chapter we will examine the neuron from a component analysis point of view. How the neuron supports the brain's requirements and the capability of neurons to support synchronous activity will be explored. We will also examine how your brain developed from a few embryonic cells into a neural system containing billions of neurons interconnected with extreme precision. All of this accomplished with no overall schematic or wiring diagram.

Neurons and Brains Evolved Together

The neuron and the human brain that utilizes it evolved in parallel. The current level of neuron complexity and variability must have been required to support the system design requirements placed on the brain for survival. The possible neuron design space is however limited by the fact that it is a living cell with a limited design repertoire. Lets examine some neuron cellular restrictions and some brain design requirements.

Cellular restrictions start with the fact that the membrane of living cells is leaky in an electrical sense; it is porous to ions that collectively generate the internal electrical state of the neuron. This makes the cell a poor candidate for long distance propagation of precise information encoded electrically. The signal simply decays to rapidly and electrical diffusion is slow. The leaky nature of the cellular membrane also imposes a strict time limit on the impact any one electrical input signal can have on the cell.

The size of a living cell is very small and neural control is required over a body that is quite large. Living cells also have the annoying problem of dying. Better not count on any one particular neuron being around. This is not a problem in cell populations where all cells are alike and stem cells simply multiply to replace the deceased cells. With minor exceptions, neurons do not multiply in the adult and multiplying neurons would be of limited help because each neuron grows into a unique shape in support of a neural circuit. That shape is a result of the neuron's history in support of the host's neural requirements. The actual interconnections a neuron makes are not encoded in the gnome.

Brain system design requirements include neural reaction times that must be very fast if an animal is going to survive for very long in a hostile world. Slow nervous system architectures would have been quickly eliminated. Neurons must support fast reaction times over distances that are vast compared with cellular dimensions

The neuron has evolved to solve the brain's system design requirements within its cellular limitations. The slow electrical diffusion and leaky electrical characteristics are utilized only in the input dendrites. Dendritic trees are limited in size and area covered to minimize the electrical diffusion time. The leaky membrane means inputs must be correlated in time to summate and have an effect on the neuron. This has the effect of limiting the pattern detection sensitivity of the neuron to input patterns that are synchronous within the decay time window of a synaptic input electrical charge. The input section of the neuron has evolved to make a sensitive pattern recognition device tuned to large numbers of inputs synchronized in time. The trigger zone is well localized in the small area of the axonal hillock. The use of a small fixed point as the trigger zone for summation of diffusing electrical input signals makes the neuron consistent and repeatable in its response.

The cellular slow electrical diffusion and leaky electrical characteristics cannot be tolerated in the neuron's output design that must signal over large distances to support large body sizes. The neuron has evolved the regenerative axonal pulse that propagates rapidly down the axon at a constant speed to solve the slow diffusion problem. The body has evolved myelin containing insulation cells that surround the axon to solve the leaky membrane problem.

To combat cells dying, the nervous system relies on massive parallelism with its inherent redundancy. Fast neural reaction times are facilitated in two ways. The slow electrically diffusing input section is limited in size to minimize the delay effects and all neural output signal transmission is accomplished via the output axon that is very fast with constant speed.

That we have few levels of neurons in neural circuits is probably due to the fact that the first vertebrates were small and basic vertebrate body design has not changed just because vertebrates have gotten bigger. The basic body plan is just scaled up. Bones, muscles, and blood vessels increase their cell count for the required expansion and neurons have expanded their axonal length. Chemical signaling between neurons takes about one millisecond so there is a small benefit in circuit speed with minimal neurons in series.

Synchronization (The Case for Synchronization of Brain Activity)

The nature of the neuron's dendritic front end favors the recognition of input patterns that are synchronous in time. Synaptic inputs that occur farther apart in time than the short decay period of any one vote cannot possibly add at the cell body. The first vote will have decayed to zero before the second vote arrives. The total number of votes within this decay window must raise the cell body electrical state by the threshold level to cause the neuron to fire. If we think of this decay time as a sliding window, the neuron is only monitoring votes within this window.

This is important from a design point of view, designing a synchronous system is a whole lot easier than designing an asynchronous system. Synchronous systems are systems where input is sampled at a selected time. This allows everything that happens outside of the selected time to be ignored. This has the effect of filtering out irrelevant input, commonly called noise. Digital computers sample their input at regular intervals with a clock. All digital computers have clocks to synchronize their activity and their performance is relative to their clock cycle speed.

But we can't build a computer out of neurons, so why is synchronization important for neurons? Remember our football stadium analogy with 60,000 people jumping up and down. Lets assume that one of the input patterns involves 10,000 people who stand up together at the same time to form a loose circle pattern. The problem is that this pattern is very difficult to recognize because of the random activity of the other 50,000 people. Now lets assume that it is night and that we turn out the lights. Every second we take a flash picture of the audience and examine it. If the flash picture occurs while the 10,000 people are standing, the circle pattern is easily recognized. The random activity between flash pictures is filtered out by the dark. It is this ability to filter out noise that makes synchronous systems simpler.

Now lets assume that somehow the flash camera is designed to always go off when these 10,000 people stand up. That would make the recognition of this pattern easy and more importantly repeatable. We still have a stadium of 60,000 people jumping up and down but someone tells us when to look. This is the equivalent of asking the person in the chair in our neuron stadium to examine still photos of the crowd that are taken when a pattern may be there. Don't bother looking unless I tell you to look. This is a neuron mechanism that would make having 60,000 inputs make sense. The neuron has evolved to support just such a mechanism.

There is a special type of synapse that is common in the dendritic spines of learning type neurons called N–Methyl–D–Aspartate (NMDA). This NMDA synapse supports the type of neural pattern sampling we have just described. In a NMDA synapse a magnesium ion (Mg+) resides in the input ion channel and blocks the channel. In the blocked state, the receipt of neurotransmitter to the NMDA synapse can affect the ion channel but no ions can flow due to the magnesium plug. The synaptic membrane must be depolarized by a secondary excitatory synapse to remove the magnesium plug before the neurotransmitter can have an input effect on the neuron.

The dendritic arbors of learning type neurons typically have two types of synaptic inputs. The first type is an excitatory input that serves to depolarize the dendritic arbor. This depolarization serves to remove the magnesium plugs of NMDA synapses and allows them to behave like a standard synapse.

Lets go over that again. The dendritic arbors of learning type neurons contain NMDA type synapses and secondary excitatory synapses. An input to the secondary excitatory synapse that removes the magnesium plug must be received before the normal voting type signal can be recognized. If this "remove your magnesium plug" signal is received by all of the neuron's synaptic spines at the same time, the equivalent of our flash photo has occurred. Only patterns that occur while the "remove your magnesium plugs" signal is active, can cause the neuron to fire. This allows the neuron to sample its 60,000 inputs only when patterns are active and filters out all of the intervening noise.

The evidence supporting the idea that learning type neurons receive an input signal that depolarizes their dendritic tree and synchronizes their pattern detection function is circumstantial. From a design perspective, the ability to gate input is extremely useful. It allows the neuron component to be controlled to improve pattern detection and filter out extraneous noise. In our analysis of how your brain works, this NMDA component feature of your learning type neurons plays a very key role in analyzing how your brain functions.

This NMDA capability in learning neurons allows signal input to synapses to be gated. This allows, with the right design, the pattern detection of the neuron to only be enabled when valid patterns exist to be detected. Gating input would also cause learning type neurons to be normally silent, which they are, and not contribute more noise. The brain utilizes NMDA type synapses on the dendritic spines of learning type neurons.

Neural Development (How You Grew Your Brain)

How the human nervous system develops is not only amazing and interesting, it also provides clues as to how the finished product functions. The development of the human nervous system in the growing embryo is just the first stage of continuous change that persists for the life of the individual. Many of the mechanisms characteristic of the developing nervous system are indicative of the processes that operate throughout life.

Before we get into the details, lets examine the basic growth plan that produces your brain. Approximately two weeks after conception, the embryo grows a layer of cells called a neural plate that curls up to form a neural tube and then grows another layer of cells that separate from the neural tube to form a structure called the neural crest. The entire peripheral nervous system (PNS) will develop from this neural crest and the entire central nervous system (CNS) will develop from the neural tube.

Neural crest cells divide to produce immature PNS neurons that migrate to various places in the body and grow into neurons relevant to that position. If you pluck one of these immature neurons headed for your leg and stick it into your stomach, it will grow into a neuron that looks and works like all the other neurons in your stomach. The immature CNS neurons produced by cell division in the neural tube work much differently. New neurons are produced in the inner portion of the neural tube and then migrate outward to form CNS structures. The function of these immature neurons is fixed at birth. They migrate to a specific location and perform a specific function. Plucking one of these neurons and placing it somewhere other than where it is intended to go will cause it to not be wired properly and just die.

The typical behavior of an immature CNS neuron is to migrate to its specified location, grow a dendritic tree structure, and then grow its axonal structure to a specified location and connect with all dendritic structures located there. This phase of embryonic development produces a vast number of neurons that all migrate, build dendritic structures, and then send their axons out and connect with everything in sight. The embryo produces many more neurons than can possibly be used and every possible wiring solution that can be built with them.

The result of all of this overproduction of neurons and wiring is a central nervous system that contains every possible neural circuit that could possibly be needed. Now the problem becomes, what circuits do you need and which ones do you throw away. Remember that synapses and neurons that are not part of a useful neural circuit atrophy and die. That neuron characteristic is utilized to eliminate all non-useful circuits. As you learn, some neural circuits become more used and others become less used and atrophy. The final result is a brain that is precisely constructed and precision wired. Lets examine in more detail how this process of growing a ball of yarn and pruning it into a diamond is actually accomplished.

Neural crest cells form all ganglion neurons, all sensory neurons and all neurons that make up the PNS visceral nervous system. These cells are born in the central location of the neural crest and then migrate to often remote body areas. The final neural form of these neurons, including the neurotransmitter they employ, is determined by the local cellular interaction that takes place at their final destination and is not a function of where in the neural crest they are born or when they are born. The neural crest neuron program after birth is migrate to this location where you will receive instructions specific to your function.

CNS neurons are born through cell division in the internal layers of the neural tube and then migrate outward to their final position as the neural tube essentially expands to become the CNS. The final resting place of these neurons is determined strictly by when and where they are born. Each successive generation has a distinct role to play and that role may not be altered.

The growing cell mass of the neural tube forms two distinct areas called the alar plate and the basal plate. These areas are separate and distinct. The back alar plate will form the input areas of the CNS and the frontal basal plate will form the output areas. This functional differentiation of dorsal alar cells developing input structures and ventral basal cells developing output structures is generally true for the entire CNS with few exceptions. It is therefore insightful to note which structures develop from which plate at various levels of the neural tube.

In the spinal cord the alar plate is dorsal and develops into thin sheets or layers of neurons with different input functions and the basal plate is ventral and develops into columns for support of output functions. In the brainstem the alar plate is lateral to the basal plate in support of cranial nerves and the basal plate is medial and forms the reticular formation that outputs to large areas of the brain. The cerebellum develops from the alar plate of the brainstem and is an input component. The inferior olivary nuclei develop from the basal plate and are output components.

The cells of the head end of the neural tube develop into all of the structures above the brain stem. The thalamus develops from the alar plate and performs an input function. The hypothalamus develops from the basal plate and is an output structure. The entire cerebral cortex develops from the alar plate and is an input component. The basal ganglia nuclei develop from the basal plate and are output components.

Two aspects of neuron development are key in realizing the final form of your nervous system. One is the migration of the neuron cell body to its final destination and the second is the growth of the neuron's axon to its target destination. For the most part axonal growth follows cell body migration but there are many exceptions where the axon extends first or the cell leaves a trailing structure during its migration that becomes the axon. The granule cells of the cerebellum are such an exception. We will limit our discussion here to the more common cell migration followed by axonal migration.

Chemical messengers that diffuse from certain locations in the body are believed to facilitate the migration of neuron cell bodies and axons. The concentration of these chemical messengers is used to identify a path and a final destination point within the body. We do not actually know the detailed mechanisms that support these precise migrations. In the brain, glial cells also provide a ladder network to guide and enable the neurons to their final destinations. Whether the glial cells are just structural aids or road maps is not known.

Immature neurons called neuroblasts are created when neuroblast cells divide near the internal layers of the neural tube. The final destination of these cell bodies is usually a nucleus being developed from that particular portion of the neural tube. The final arrangement of neurons within a nucleus is a precise matrix based on the neuron's birth date and birth location. Their dendritic trees will also be an overlapping reflection of that same matrix. The axonal growth paths of these neurons will maintain the same matrix positions as the cell bodies. Their axonal endpoints and collaterals will be an overlapping reflection of that same matrix much like their dendritic structures.

Following cell body migration the neuron grows its dendritic structure in a fixed pattern in the area surrounding the cell body and begins axonal growth. An axonal growth structure called a growth cone facilitates the migration of the growing axon to its destination. The growth cone constantly sends out multiple filoplia that are rope like structures that pull and guide the growing axon. Cut off the growth cone and the axon stops growing. The migration of many axons involves multiple growth programs. The axon will migrate to a location containing a guidepost cell and then stop growing for up to 24 hours. Then the axon will continue growth until a new destination is reached.

The growing axon sends out collaterals along its path to target locations and stops growing when it reaches its target destination. The neuron then puts out many collateral axonal end points on all of the dendrites available in the destination area. This results in a great many synapses being formed that will later be pruned.

The migration of the neuron cell body and growth of the axon are driven by a basic migration program that is designed to place the neuron cell body and axonal end point in precise locations. The growth of dendrites and axonal collaterals are pattern driven and are not precise at all. The actual final wiring solutions will be tuned by actual usage. This growth plan produces a neural wiring solution that is very precise in its overall layout but highly variable in its interconnections. The gnome specifies the location of neurons, their dendritic shape, and their axonal target location.

Note that the chemical markers that determine a neuron's location do not have to specify the destination precisely. They only have to specify the location of the nucleus that the neuron is a part of. All of the neurons that make up the nucleus can start with the same target destination. The neuron aligns itself with its birth mates. This greatly decreases the amount of information that needs to be specified in the genome. It also allows that information to remain the same as body types evolve and nuclei grow to keep pace.

Almost all input and output paths cross the midline at the top of the spinal cord and cause the right side of the brain to control the left side of the body and vice versa. The most primitive neural tracts that are made up of unmyelinated small axons conducting pain input, tend not to cross the midline. The first crossing seems to have appeared in conjunction with the brain stem. All animals of any complexity display this right to left characteristic.

A conspicuous problem that this crossing solves is the disparity between the size of the brain and the size of the body the brain controls. Migration based on a chemical marker results in precision getting worse with distance. If a chemical marker has a concentration of 0 to 100 and a migrating neuron can accurately locate to one unit of chemical concentration, one hundred possible location points within one foot is much more precise than within ten feet. In a dinosaur one set of spatial markers is not going to work very well. A solution is to implement multiple copies of the same axonal migration program, one migration for the large body and one that is much more accurate for the smaller brain.

The development paradigm just articulated produces neurons in precise body locations, in precise alignment relative to neighboring cells, precise axonal end points, and many random connections between all neurons. The development scheme ensures that all needed neural circuitry exists by producing many more neurons than are required. This coupled with the overproduction of synaptic interconnections results in a vastly over wired brain in the new born infant. In fact, the growth of synapses continues for months after birth. This over wired brain is trimmed in both neurons and interconnections by the use it or lose it law of neurons. The development of the brain following birth can be thought of as neuronal survival of the fittest. Neurons compete to be part of the final wiring solution of the brain. Many fail and die.

The greatest number of synapses a person has in their lifetime is around eight months old. After that it is all downhill. The visual cortex declines from approximately 3500 billion synapses at age eight months to just over 2000 billion synapses at age eleven. That is a loss of over 40% during a period when your eyesight and ability to recognize objects is getting substantially better by all measures.

Some parts of the brain will lose up to 80% of their neurons between birth and adulthood. In early development, muscle cells are innervated by multiple motor neurons. Only the synapses from the one motor neuron that is the most efficient will survive on each muscle cell. Motor neurons support approximately 10,000 dendritic synapses. One motor neuron will achieve dominance over all others in pattern detecting this complicated input from multiple sources for each muscle cell. This results in many synapses atrophying and motor neurons dying in a final fine tuning of the motor neuron to muscle wiring. The cerebral cortex will lose approximately 40% of its neurons before adulthood and very few after adulthood.

The loss of neurons in the brain does not stop with adulthood but continues throughout life. Some areas of the brain such as the brainstem, hypothalamus and cerebral cortex lose relatively few neurons with advancing age. The hippocampus and substantia negra lose around 25% and 50% of their neurons respectively by the age of 75 in normal humans. The weight of the aging brain also falls with age and a normal adult will lose approximately 25% of brain mass between 20 and 80 years of age. You lose approximately one million neurons a day for every day of your life. It is only now in modern times that neuron loss with age results in loss of quality of life. With the average life span greater than the mid seventies, some detectable age related dementia and senility are now commonplace for the first time in the human species.

The growth of your brain in the embryo and after birth necessitates the pruning of unneeded neurons and synapses to realize the final result. The use it or loose it nature of the neuron is what enables this over growth and pruning process to work. Synapses and neurons that are part of a useful neural circuit are strengthened and grow. Synapses and neurons that are not part of a useful circuit atrophy. The synaptic meaning of useful is that the synapse opening causes the neuron to fire, remember our Mr. Hebb.

An input signal to a synapse must be synchronous in time with enough other inputs to cause the neuron to fire. Random input is unlikely to sum to the firing threshold level of the neuron. Input that is synchronous will self select neural circuits that propagate it most efficiently because these circuits will fire more readily and be strengthened. Neural circuits that are less efficient at recognition of this synchronous input will atrophy. The growing brain retains all neurons and neural circuits that receive and propagate synchronous input. The growing brain retains all neurons and neural circuits that are useful.

A great example of precision wiring in the brain is the ocular dominance columns located within input layer four of the primary visual cortex (V1). The two dimensional visual field of each eye is broken into small patches and interspersed with the other eye. The entire layer four of the primary visual cortex can be thought of as a perfect checkerboard with the left eye one board color and the right eye the other.

All of the neurons involved, the retina, thalamus, and cerebral cortex, are born separately and migrate to their destinations. Each builds its relevant dendritic tree and extends its axon. The ganglion retinal neuron sends its axon to the thalamic lateral geniculate nucleus and forms synapses over a wide area. The lateral geniculate nucleus neuron sends its axon to the primary visual cortex and synapses over an area that will cover several ocular dominance columns in the final design. The makings of the visual input system are now in place. There are more neurons and many more synapses than required for the final design.

Electrical gap junctions that interconnect the cells of each retina synchronize the firing of retinal cells. In the embryo each retina begins to fire spontaneous bursts of action potentials very early in development. These synchronous bursts from each eye of the embryo last for about two seconds and fire about every two minutes.

The neurons in the thalamic lateral geniculate nucleus are in a competition to be the most efficient at pattern detecting this synchronous input and passing it on to the primary visual cortex. This visual input is made up of four kinds of signals, on and off P type and on and off M type retinal cells. The axonal growth paths of these four input types for each eye concentrates into a slightly different area of the lateral geniculate nucleus. These different areas become more and more sharply defined as the competition unfolds in the developing embryo. The synchronous input from each retina results in a segregation of the inputs from each eye into precise layers of neurons in the lateral geniculate nucleus.

Development generated a diffuse set of wiring that was guaranteed to cover the required solution and then pared that wiring solution to its final form with artificial synchronous use. The firing lateral geniculate neurons are having the same effect on the primary visual cortex during this time. As the lateral geniculate neurons are forming precise areas for these four visual input paths, their synchronous projections to the primary visual cortex are causing the neurons of layer IV to form precise areas of alternating input from the two eyes called ocular dominance columns. At birth the lateral geniculate nucleus is completely wired in its final form and the checkerboard of ocular dominance columns in the primary visual cortex is roughly formed but not complete. The final segregation per eye in the primary visual cortex is accomplished through actual visual experience.

This wiring of ocular dominance columns is never really over. Lazy eye, a condition in some children, is a result of one eye gaining dominance over the other in the primary visual cortex. The ocular dominance columns of one eye grow and shrink the area available to the other. This can be fixed easily with an eye patch to allow the lazy eye to reclaim its lost territory. People blinded in one eye with no retinal input from that eye will convert their entire visual sensory input layer to the remaining eye. This example of the formation of ocular dominance columns is not unique within the developing nervous system. It is typical of the birth, growth, over wiring and subsequent pruning for final design that is a part of the development of the entire nervous system.

The wiring of the primary visual system is also a very good example of what happens in the human brain as a result of injury. The axonal connections of a neuron in the human brain are extensive. The wiring remains somewhat diffuse even after the paring of neurons and synapses with use. When a part of the brain is damaged by stroke or injury, the synapses that are no longer part of an operational neural circuit atrophy and disappear. Any and all synapses that do continue to carry useful information concerning the functions affected by the physical damage will be strengthened and increased. This can result in significant rewiring of neural circuits to compensate for the injury. This is the result of the basic neuron's standard function and not some change brought about by the injury. Each of your neuron bricks is a stand-alone cell operating by a standard set of principles that is not altered by age.

From birth to adulthood your brain continues to grow and create wiring to accomplish the functional demands you place on it. The brain increases its weight about four times during this period as the individual surviving neurons grow and create synapses. The brain becomes less malleable as you grow to adulthood. The wiring necessary to perform some functions must be formed during this developing period of your life prior to adulthood or it will never be formed. Language and mathematics are examples of capabilities that must be taught and learned when the developing brain is still in its plastic wiring state of growth. People not taught to speak a language prior to becoming an adult, never learn to speak properly. This is due to the complexity of these capabilities and the amount of wiring necessary to perform them. This amount of new neural wiring cannot be created in the adult brain. They say you can't teach an old dog new tricks, it's true.

The growth and expansion of each particular area of the neural tube varies depending on the structures being grown from that portion of the neural tube. Dividing spinal cord neuroblasts push older cells outward to form successive internal layers of cells. The back alar cells form layers of cells with similar input functions in the dorsal horn and the frontal basal cells form columns of cells with similar output functions. The lower brainstem develops much like the spinal cord. The upper brainstem alar cells grow to form the cerebellum. The cerebral cortex neuroblasts actually migrate outward to form successively more external layers of the cortex. The internal layer six is formed before layer five and so on and the cells for each layer must migrate past the layers already in place.

The oldest portions of the cerebral cortex are the first of that structure to appear, the hippocampus and the cingulate gyrus. The growth of the cerebral cortex is like the blowing up of a balloon except the material is not stretched but is increased in area by dividing cells. The beginning of the frontal cortex is the first area to be added followed by the parietal cortex, the occipital cortex and finally the temporal lobes. At six months the embryonic cerebral cortex still resembles a smooth balloon. After this time the expanding cortex runs out of room and begins the folding process that gives it its final appearance. The axonal connections within the cortex affect which areas become sulci and which gyri. Since the cortical wiring targets are specified in the gnome, humans all have similar major groves or sulci patterns in their brain.

At birth, the outside layer one of the cerebral cortex is fully populated with neurons that are wired and ready for use. Layers two through six are populated with neurons but continue to form synaptic wiring for many months. These layers are not ready for use and layer one serves as the newborn's cerebral cortex at birth. Over time, layers two through six mature and begin to take over the standard cerebral cortex function in the young child. Layer one atrophies as the other layers take over its function. In the adult, layer one contains the upper portions of the dendritic trees of pyramidal neurons in layers beneath it and virtually none of its original neurons.

The growth of the heavily interconnected components that enable human intelligence, the thalamus, cerebral cortex, and basal ganglia, is instructional. An area of declarative cortex reciprocally connects with an area of frontal cortex. Each of those two interconnected cortical areas projects input to one area of the basal ganglia. They each reciprocally connect with separate gating nuclei in the thalamus. That same basal ganglia area projects to the thalamic gating nuclei supporting the frontal cortex area. All of these interconnected neurons in the cerebral cortex, thalamus and basal ganglia are born at the same time in the growing embryo. The development program for interconnecting neurons of the cerebral cortex and supporting neurons of the basal ganglia and thalamus seems to be, migrate to your specified location, align yourself with your birth mates, and project to neurons that were born at the same time as you. Another example of a simple yet precise solution that minimizes the amount of genetic information required and supports the expansion and growth of these structures.

Conclusion

Now lets take what we know about the component specification of the neuron and match it to the system design requirements of the human brain. The neuron cell body does not act like a person in a chair in a stadium. It really is just the central point for collection of passively diffusing electrical potentials from all inputs. There is absolutely no evidence that it is anything else. It is the target in many cases for inhibitory input because of its proximity to the axonal hillock trigger zone. It is in the cell body that cellular changes are signaled to the gnome and implemented in protein production. The cell body is the command center for cellular change and structural integrity as it is in all cells. The neuron's axon for all of its specialty in action potential generation and transmission is also a passive device. Its function does not change under any circumstances. It is only the copper wiring of the brain's circuits.

It is in the dendrites where all the action is. This is the area of the neuron that supports the changes necessary to enable the plastic capabilities of your brain. Learning and memory are implemented by the acquisition and subsequent recognition of patterns due to changes in the dendritic arbors of learning type neurons. The patterns recognized by hardwired neurons do not change and they each make up a small part of a static neural circuit.

When a learning type neuron fires, the synapses that are active are slightly enhanced. A single learning type neuron is a small part of a forming memory recognition. Memory recognition, such as recognizing previously viewed pictures, most likely involves large areas of declarative cortex and is made up of hundreds of millions, if not billions of synapses. A very small individual synaptic enhancement over very large numbers of synapses adds up to a large change overall. Enough to allow you to easily recognize previously viewed material.

Information flow in the brain is continuous and massively parallel. Each pattern recognition neuron fires and recovers to signal recognition again. Memory, learning, and intelligence are all built from circuits containing neurons that perform pattern recognition. Nothing is stored in the brain as information in a computer sense.

Neurons are amazingly quick at altering their input recognition circuitry. You can recognize a building, a painting or a face that you have only seen once after a great deal of time. You may not remember when or where you saw a face but you know you have seen it before. As you drive somewhere you have been once years before, you can recognize scenes from the previous single trip.

Neuron pattern detection capabilities are sometimes altered between different states. The neurons of the basal ganglia striatum enter an up state when they are activated by large numbers of highly synchronous excitatory inputs from the cerebral cortex. They actually enhance and hold their ability to fire in the up state by changing their internal voltage. In the down state these neurons are very hard to excite. It takes a very large set of highly synchronous inputs to get the neuron to fire. The result is a neuron that is very particular about signaling pattern detection but once it does, it will continue to fire with a lot less input.

The brain produces as many neurons as can fit in the space available during development and wires them together. All of the neurons that are useful survive. This development algorithm actually gives you the maximum number of possible pattern recognition devices that can possibly be used to perform the functions required. Parallelism and redundancy are the outcome of this process of maximum growth followed by pruning.

Now our picture of your neuronal brick is complete. Learning and memory are enabled because synapses that are active when the neuron fires, are strengthened. This Hebbian principle is what allows the learning of pattern recognition. Neurons that fire in recognition of patterns grow their dendritic structure to improve their sensitivity to those patterns. This coupled with the atrophy of non-useful neural circuitry enables the plasticity of the learning portion of your brain that serves you throughout your life.

The nature of the neuron's dendritic tree causes the neuron to detect input patterns that are synchronous in time. The neurons in the learning portions of your brain have evolved NMDA type synapses that allow the gating of input patterns. This neuronal brick remains as our only building material for the construction of the human brain; it is up to the task.

Chapter 5

Human Brain Components Functional Analysis

Introduction

The goal of this chapter is to examine each of the various neural components in enough detail to support analysis. Not just circuit analysis, functional analysis. We are going to cover all brain subsystem components in the same order as the brain building tour. We will again begin with neural input and finish with neural output.

The organization of component information will be consistent; a physical description of the component, where it resides in the physical brain, embryonic development of the component, when the component first appears, the component's internal neural wiring, all connections to other components, and neural systems in which the component plays a role. Finally, based on the information presented, we will discuss the component's function and how the component performs that function.

Neural Input

Your perception of your world is constructed from sensory input. Humans rely mostly on visual input but auditory, somatosensory, smell and taste inputs also contribute to your perception of reality. Each of these inputs is pattern detected into its constitute parts and further pattern detected to build your virtual world.

All of your neural input systems exhibit the same generic characteristics of linearity, inhibition, thalamic gating and dedicated primary cortex. Input signals representing the external world are ordered within the CNS as a precise linear array throughout the entire input path, including their representation in primary sensory input cortex.

Segregation of input information within this strictly maintained linearity occurs in two ways. Incoming data is segregated by receptor type. The P and M type ganglion cells of the retina and the various somatosensory receptors are prime examples. These different types of sensory information take separate physical input paths, are gated by different neuronal layers in thalamic nuclei and are received by different portions of primary sensory input cortex. Secondly, the receptor type density and resolution is maintained in the entire input path including the primary sensory input cortex. It is this attribute that results in the distorted body map in the S1 cortex and the large area dedicated to the fovea in V1.

Inhibition is pervasive in all input paths except vision. Inhibition at each neuron level of the input path comes from two sources. Collaterals from neurons in input paths inhibit adjacent input path neurons. This facilitates the ability to precisely discriminate the location of a strong input signal within the two dimensional array. Also, the cerebral cortex controls inhibitory signals to all levels of input paths. Local inhibition between input neurons contributes to what gets attention but the cerebral cortex maintains final authority. This enables the cerebral cortex to focus attention on a subset of incoming information as both a neural reflex and a conscious decision.

The thalamus is the switchboard for all incoming sensory information except olfactory. Like a switchboard, the thalamus is a hardwired device. The segregation of input information by receptor type is physically maintained through thalamic input circuits. Cerebral inhibition control is the most extensive and most refined in the thalamus.

The various primary sensory input cortices share some common characteristics. Sensory input remains linearly mapped and segregated into vertical stellate neuron rich granular columns. There is limited overlap between the receptive fields of columns that collectively represent the entire input array. The actual content of the input array in each column remains plastic throughout life. Behavioral demands for increased resolution within an area of an input array results in the primary sensory cortex providing finer resolution for that area. The representation across cerebral granular columns shifts to provide more columns dedicated to the area in need of finer resolution. The ocular dominance columns of the visual cortex, the precise audio abilities of a piano tuner and the heightened finger sensitivity of a Braille reader are examples.

The functions of primary sensory input cortex are to select and control through inhibition the specific input made available for attention, to dissect that input through pattern detection, and to make the dissected input available to secondary areas of cortex for higher levels of pattern detection.

Somatosensory Input (How You Feel Your Environment)

Somatosensory input provides the brain with a full body representation of the skin's interface to the external world and the status of all joints and muscles that make up the body. Many specialized sensory neurons act as somatosensory receptors and detect different types of tactile information over different time spans. The receptor density in body areas determines your tactile sensitivity for that body part. The highest concentration of receptors occurs in the fingers of the hand and in the tongue. There are approximately 2500 receptors per square centimeter in the fingers of the hand providing input to approximately 300 receptor neurons. Somatosensory receptive fields in the lower body trunk are 100 times larger than in the fingers of the hand. The neural input path for all receptor types is segregated for the entire input path including the receiving column in the primary sensory cortex.

The spinal cord or brain stem houses the secondary neuron in the somatosensory input path. Inhibition of input paths at secondary spinal neurons is controlled by pyramidal neurons in layer five of the primary sensory cortex (S1). The reticular formation receives collateral input from all input tracts on their way to the thalamus. Those collateral sensory inputs to the - reticular formation are a prerequisite to recognition of that input. Many input signals carrying proprioceptive information leave the brainstem to enter the cerebellum.

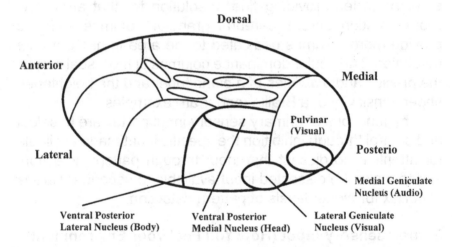

Figure 5-1: Sensory Input Thalamic Nuclei

The thalamic nuclei that receive and distribute somatosensory information are the ventral posterior nuclei. (Ref. Fig. 5-1) These nuclei are the main switchboard components for all tactile information bound for the primary sensory cortex (S1) and they also signal to the secondary sensory cortex (S2). The medial nucleus handles information concerning the face and head and the lateral nucleus handles the body.

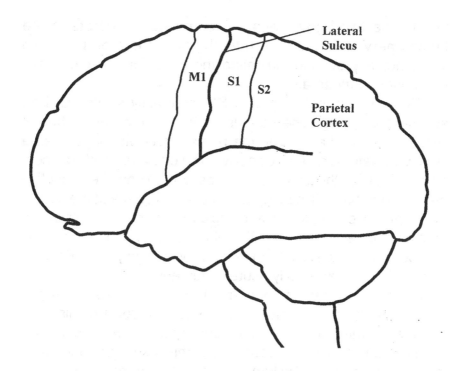

Figure 5-2: Somatosensory Cortex

The primary sensory cortex (S1) is a narrow strip at the front of the parietal cortex. (Ref. Fig. 5-2) Just to the rear of S1 is the secondary sensory cortex (S2). S1 contains a full body map of somatosomatic input that is similar in all vertebrates. Like all primary input cortex, S1 contains discrete stellate neuron rich granular columns that receive the sensory input via the thalamus. These granular columns are surrounded by stellate poor, pyramidal neuron rich dysgranular type cortex.

Each granular column in S1 represents the input from a continuous and discrete region of the body. The body is completely represented by a mosaic of discrete body area patches in the granular columns. Each discrete column receives input derived from one specific type of receptor neuron in the periphery of the body. These different types of input actually interface into different sections of S1.

A mirror image motor map in the primary motor cortex M1 in the frontal cortex closely matches this precise body image in S1. The two body maps are completely separate in the human nervous system. In rats, the paw areas of these two maps actually overlap indicating the transitional evolutionary separation of these areas.

The neurons that comprise S1 are mostly silent with little spontaneous firing. The various cortical layers have different receptive fields. Layer four, the primary receiving layer, has a small well-defined receptive field with limited overlap with adjacent areas. Each of the other layers has a more diffuse receptive area. Cortical layer five projects to the spinal cord and brain stem for the purpose of inhibition and has a much larger receptive field than any of the other layers. The ability to discriminate precisely the location of tactile information is most highly developed in primates and is especially acute in humans.

Proprioceptive input concerning the status of muscles and tendons is provided mainly to the cerebellum. Some of this input does reach the primary somatosensory cortex S1 and you do have limited conscious awareness of this input. Proprioceptive input enters the cerebellum in either the inferior or superior cerebellar peduncles.

The body's interface to the exterior world is projected to the contralateral thalamus and on to somatosensory cortex. (Ref. Fig. 5-3) Note that all of these inputs from both the body and head provide collateral axonal connections with both the reticular formation in the brainstem and the intralaminar nuclei in the thalamus, in order to be recognized.

Somatosensory inputs are projected to S1, S2, the posterior parietal lobe, and directly to the primary and secondary motor cortices. Input to the dysgranular matrix of S1 comes from neighboring granular columns, secondary sensory cortex S2, the contralateral same area of S1, and from the primary and secondary motor cortices M1 and M2.

The primary sensory cortex S1 is further divided into four subzones: 3a, 3b, 1 and 2 (Ref. Fig. 5-4). There are actually four separate full body maps contained in the granular columns of these sub zones and a fifth in S2. Most of the input to the primary sensory cortex is directed to subzones 3a and 3b with

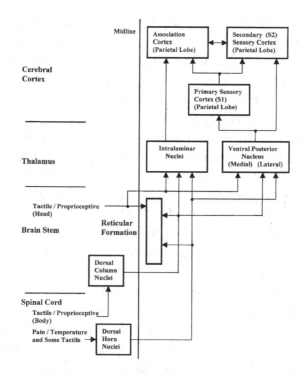

Figure 5-3: Somatosensory Input System

more sparse input to 1, 2 and S2. There is no corresponding body map that can be detected in the dysgranular matrix surrounding the granular columns.

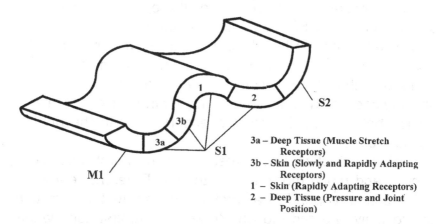

3a – Deep Tissue (Muscle Stretch Receptors)
3b – Skin (Slowly and Rapidly Adapting Receptors)
1 – Skin (Rapidly Adapting Receptors)
2 – Deep Tissue (Pressure and Joint Position)

Figure 5-4: Primary Somatosensory Cortex

Each of these S1 subzones has a different dominant input type. The dominant input type to 3a is proprioceptive input from muscle stretch receptors and this zone is concerned with position sense only. The columns of 3a have small receptive fields that in the hand cover one or two phalanges of a finger. The 3b dominant input type is cutaneous tactile receptors and 3b is concerned with touch only. Again the receptive fields for the columns of 3b are quite small, analogous to the receptive fields of 3a. The major input type to zone 1 is rapidly adapting cutaneous receptors concerned with touch. This zone is heavily concerned with the texture of objects. The receptive fields of zone 1 are somewhat larger, typically several fingers of the hand. Zone 2 input is predominantly deep pressure receptors and this zone is concerned with both touch and position sense. The receptive fields of zone 2 are like that of zone 1 and this zone in the hand is concerned with the size and shape of objects.

All of these subzones are reciprocally interconnected with each other. Sub zones 3a and 3b project to the secondary sensory cortex S2 in addition to zones 1 and 2. Zone 2 projects to the rest of the parietal cortex and also to the primary motor cortex. If the path from zone 2 to the primary motor cortex is cut in monkeys, the animals are unable to pick up objects with their hand. Direct feedback of somatosensory input to motor output is required for the correct operation of precise motor tasks.

S1 cortical layers two and three interconnect with secondary cortex S2, posterior parietal cortex, motor cortex and contralateral primary cortex S1. Layer five provides inhibitory control projections to the brain stem and spinal cord and also signals to the basal ganglia and pontine nuclei. Layer six provides return projections to the ventral posterior thalamic nuclei that provide the somatosensory input to the primary sensory cortex.

The sense of motion across the skin is first detected in zone 1 and much more fully in zone 2. Different neurons in those zones respond to all motion, gross directional motion, or precisely oriented motion. Zone 2 is the first zone where discrimination of three-dimensional objects takes place.

The granular columns of each subzone receive direct input from the body's periphery divided into a fine-grained map. The dysgranular area surrounding the columns integrates this primary input information with motor information directed to the same area, information concerning input from the contralateral body area, and feedback information from S2 concerning the area.

The flow of pattern detection starts in 3a with where the body is. This information is combined with what the body feels in 3b to provide a small-grained set of input to the higher levels. Zone 1 combines this input over a larger receptive field with primary touch input to integrate the texture of the body's interface with the real world. Zone 2 integrates all of this with large-scale body position information to discern what type of objects you are touching over all of your body parts. Closing your eyes and identifying objects with your hands represents the highest level of somatosensory pattern detection and integration of that sense with the other areas of the brain. A Braille reader's ability to read with their fingers is a great example.

Visual Input (How You See Your World)

You rely on vision as your primary input to discern the state of your immediate environment. The visual capabilities of the human primate have evolved to tower over the visual abilities of all other mammals. As opposed to most mammals, your visual input system allows you to detect colors and textures and to discern stationary objects with ease. Even though your visual system is more highly evolved, you share much of the same neural circuitry with all mammals for basic visual control. The areas that support vision are the most researched areas of your brain. A very large amount of information is available concerning the neural system design supporting human vision and we know a great deal about how you visually perceive the world around you.

Your visual system builds a dynamic three-dimensional representation of the world from a series of static two-dimensional snap shots provided by the retinas of your eyes. Your eyes move to new locations in the environment by movements called saccades. Pick a point around you, focus on it and then try to move your eyes smoothly to another point. Your eyes automatically perform saccades and focus on intermediate points. Saccades are pervasive to the way that you see. Your visual cortex, through perceptual filtering, totally ignores all visual incoming information during saccades.

The neural components that support human vision are the retina of the eye, the superior colliculus, the thalamus and visual cortical areas. We will begin with a quick review of the retinal neurons and their function. First-order visual receptor neurons are either rods or cones and they detect photons that fall upon them. There are three types of secondary neurons that interconnect first order receptors to third order neuron ganglion cells. Ganglion cells support the pattern detection of circles with a central dot and an antagonistic surround.

On center ganglion cells detect a light dot with a dark surround; off center ganglion cells detect a dark dot with a light surround. The surround input is higher priority than the dot input and a negative surround will cancel out a positive dot input. Ganglion cells are never silent and their output is modulated up or down depending on their input match for their circle detection pattern. Ganglion cells also fire to indicate abrupt changes in their detected pattern to support the detection of movement. Each ganglion cell signals continuously the closeness of their input pattern to their optimum circle pattern and also changes to or away from their preferred circular pattern.

There are two types of ganglion cells and all rods and cones supply inputs to both types. M type ganglion cells have large receptive fields and detect motion. P type ganglion cells have small receptive fields, are more numerous than M type cells, and support fine detail and color vision. There are four parallel vision input paths and each provides input from the entire retinal visual field. There are on and off center M type paths and on and off center P type paths. These paths are separate and distinct through the thalamus to the primary visual cortex.

The midbrain superior colliculus is the top of several layers of neural structure that make up the colliculus. Within those layers are neural input and structure to support four separate maps of the environment, visual, auditory, somatosensory, and an output motor map. The colliculus contains these maps of the immediate environment in order to control eye related movements. The colliculus completely controls reflex movements of the eyes, head and neck to bring the eyes onto startling inputs and under cerebral cortex influence, controls the visual tracking of movement and eye saccades.

Regions of space in the environment are represented in the colliculus much like columns in the cerebral cortex. Sensory maps and a motor map concerning a region of space are vertically aligned in layers. The motor map is contained in the deepest layer of the colliculus and projects to motor neurons to facilitate reflex movements of the neck and head. This motor map output also projects to the pontine nuclei and on to the cerebellum.

The superior colliculus provides many autonomic controls that support vision in all mammals. Some examples are the purely reflexive pupillary dilation, constriction of ciliary muscles restricting the amount of light energy allowed into the eye and the shape of the eye lens. Retinal ganglion connections to both right and left midbrain nuclei explain why a light shown in one eye causes both eye pupils to constrict.

The superior colliculus represents a highly evolved visual control system common to all mammals. In the human, this visual control system operates in close concert with parietal visual cortex in the representation of space and with the frontal eye fields of the frontal cortex for control of saccadic eye movements.

The thalamic lateral geniculate nucleus receives input from the contralateral vision field and passes that information to the ipsilateral primary visual cortex. The geniculate nucleus consists of six distinct layers with no interconnections between the layers. (Ref. Fig 5-5) These layers are stacked vertically with layer six the top layer. Layers one and two are called magnocellular layers and consist of large cells that receive input exclusively from M type retinal ganglion cells concerned with detection of motion. Both layers project independently to the primary visual cortex.

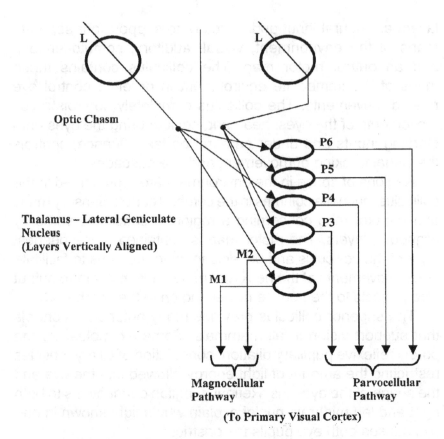

Figure 5-5: Lateral Geniculate Nucleus

Layers three through six are called parvocellular layers and consist of smaller cells that receive input exclusively from the P type retinal ganglion neurons concerned with color and fine detail. All four layers project independently to the primary visual cortex. The right lateral geniculate nucleus receives input from the left visual field and vice versa for the left lateral geniculate nucleus.

There is a precise vertical registry of the two dimensional retinal visual field maintained through all six layers of the lateral geniculate nucleus. Each neuron receives input from a small number of retinal ganglion cells. The receptive field size and on or off center circular pattern detection is maintained by the lateral geniculate nucleus and passed unaltered to the primary visual cortex. The lateral geniculate nucleus also projects to and receives reciprocal connections from areas in the secondary visual cortex surrounding the primary visual cortex.

The thalamic pulvinar nucleus located adjacent to the lateral geniculate nucleus is also dedicated to vision. (Ref. Fig. 5-1) The pulvinar receives input from the superior colliculus and receives no direct input from the optic tract. The pulvinar has reciprocal connections with secondary visual cortices and surrounding cortex of the parietal and temporal lobes involved with visual information. The spatial map constructed within the superior colliculus is provided via the pulvinar to the areas of cortex concerned with visual space, visual objects within that space and visual attention.

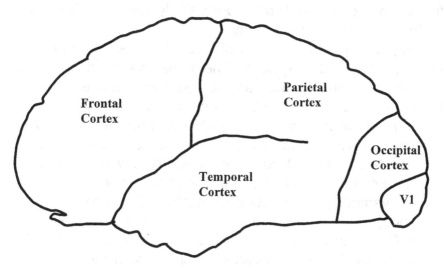

Figure 5-6: Primary Visual Cortex

The fact that vision is your main perceptual tool is reflected in the amount of cerebral cortex utilized in deciphering your visual images. All four of the major areas of the neocortex are involved in vision pattern detection or control. The main visual cortex area is the occipital cortex that is entirely devoted to the pattern detection of visual images. The occipital cortex contains the primary visual cortex (V1) at the most posterior portion of the cerebral cortex and also secondary visual cortex areas. (Ref. Fig 5-6) The inferior parietal cortex and temporal cortex contain secondary visual areas involved in the visual representation of space, motion analysis, and the recognition of objects within the visual field. The frontal cortex contains an area called the frontal eye fields that project to the superior colliculus and direct the desired location of eye saccades. These cerebral areas along with the visual thalamic nuclei and superior colliculus represent the large amount of human brain structure devoted to visual pattern detection.

Columns of neurons are organized vertically throughout the primary visual cortex. Thalamic input is received within layer four of these granular columns and fans out from there to other cerebral layers and on to secondary visual areas. The receptive fields of neurons past V1 get progressively larger and the visual pattern detected more complex.

Post V1 pattern detection of visual input is performed by two separate parallel systems. The dorsal system involves the occipital up to the inferior parietal lobe and is concerned with the representation of where. This dorsal system is involved in guiding actions such as reaching and eye movements. Many of the neurons in the dorsal system are concerned with movement. The posterior parietal cortex is concerned with space and movement and projects strongly to the frontal cortex.

The ventral system involves the occipital down to the inferior temporal cortex and is concerned with the representation of what. The ventral system is involved in object recognition. Both the dorsal and ventral systems are innervated by reciprocal connections with the pulvinar nuclei and their output projections converge on association cortex. The receptive fields of visual pattern detecting neurons tend to be larger in the right cerebral hemisphere and smaller in the left hemisphere. The right hemisphere seems somewhat dominant in representation of spatial information and the left hemisphere seems somewhat dominant in the representation of object information.

The visual input path is identical in all humans. From the circular receptive fields and concentric circular patterns of the retina and thalamus to the line orientation and rectangular receptive fields of V1 neurons, the synaptic interconnections provide each neuron in the path with precisely the required input to perform its specialized unique pattern detection function. The overproduction and subsequent paring of neurons and synapses in development always results in a precisely ordered, retinotopically mapped, similarly structured V1 in all humans. The actual size and position of V1 in humans exhibits a great deal of individual variety.

The visual system maintains linear retinotopically precise visual maps in V1 and areas of secondary cortex. There have been thirty-two such maps identified in the visual cortical systems of monkeys. The retinotopical map of the fovea occupies the most posterior portion of V1. Approximately one half of the neural mass of V1 is dedicated to processing input from the fovea and the area just around it due to the high density of receptors in the fovea.

V1 performs four distinct functions for each and every retinal area. The image is broken into line segments and separate output is provided for all line orientations within each retinal area. Output concerning color segregation within each retinal area is provided. Binocular vision or the differences between the left and right retinal images and output concerning movement perpendicular to all line orientations within each retinal area is provided.

V1 exhibits a fine structural array with both vertical and horizontal consistency. The vertical structure contains the same six cortical layers and a large layer four common to all granular cortex. The receiving stellate neurons in layer four of V1 exhibit the same circular receptive fields and on, off circular pattern detection carried through the thalamus.

The vertical layered structure of V1 is further divided into columns and blobs. Columns traverse all six layers of the cortex and consist of neurons with similar pattern sensitivities. The first type of columns is called ocular dominance columns discussed as part of neuronal development and plasticity. Inputs from each eye enter into discrete patches of V1 cortex in alternating columns of left and right input.

These ocular dominance columns are further differentiated into columns of orientation. A column of orientation contains neurons that are tuned to pattern detection of a line or edge in its receptive field that is at a certain angle. Columns of orientation servicing a particular part of the retinal receptive field are arranged in linear fashion with the sensitivity angle progressing about ten degrees between adjacent columns. A blob is a group of cells interspersed between columns of orientation that are sensitive to color within the receptive field. Blobs occupy layers two and three of the vertical structure and are cone shaped.

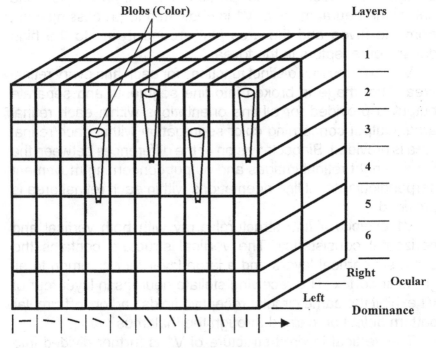

Figure 5-7: V1 Hypercolumn Structure

Ocular dominance columns, columns of orientation and blobs are arranged in an ordered array called a hypercolumn that serves a particular portion of the retinal field. (Ref. Fig. 5-7) Each area of the retinal image is serviced by a hypercolumn that pattern detects the image for that retinal area and produces output concerning line segments of various orientations, movement perpendicular to those line segments, color information and image differences between the eyes.

Retinal visual input coming into the primary visual cortex comes from the separate layers of neurons within the lateral geniculate nucleus. Alternating ocular dominance columns receive separate input from each eye. Columns of orientation receive specific on and off circular pattern information that allows them to pattern detect lines and edges of particular angles. The input connections from the lateral geniculate nucleus are wired in such a way as to shift the angle of detection about ten degrees between adjacent columns. Blobs receive input from the intralaminar nucleus of the thalamus and parvocellular input to allow the detection of color.

The primary visual cortex is a vast array of hypercolumns. Pyramidal neuron output from V1 contains vertical downward projections plus collateral horizontal projections that tie the primary visual cortex together as a hypercolumn matrix. Layers 3 and 5 send long axon collaterals horizontally that branch into clusters of axonal synapses at hypercolumn width. These horizontal connections interconnect columns within hypercolumns of similar pattern detection. In this manner orientation columns of similar angle sensitivity are tied into a cohesive whole sensitive to one particular line orientation.

Each hypercolumn is approximately one square mm and services a specific retinal field. Each individual column within the hypercolumn projects horizontally to its counterparts in adjacent hypercolumns tuned to similar pattern detection tasks. The optic tract input to this array of hypercolumns is a precisely retinotopically mapped two-dimensional image of the contralateral visual field. Like the similar linear mapped input fields in S1, the V1 retinal fields serviced by hypercolumns are basically non overlapping but can exhibit a great deal of plasticity.

V1 is precision wired to accomplish the decomposition of the retinal image. Vertical structure is concerned with specific input and output interconnections. Horizontal structure interconnects areas with the same response properties. Precise orderly structure in both the vertical and horizontal planes combined with precise repeatable modules allows the primary visual cortex to completely dissect the retinal image and output the dissected image to secondary visual areas for further pattern detection.

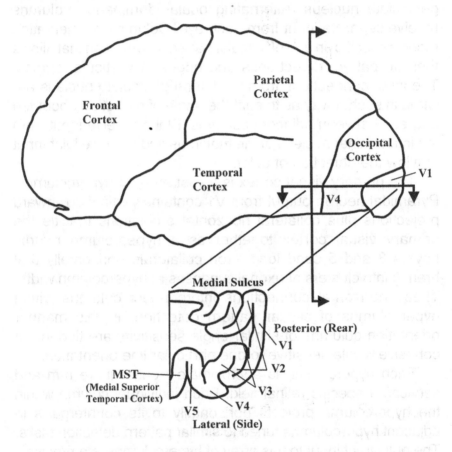

Figure 5-8: Secondary Visual Cortex

There are four secondary visual cortical areas and each of these areas receives input predominantly from V1 by direct and indirect routes. Much of secondary visual cortex is buried in the median sulcus at the back of the head that divides the left and right hemispheres (Ref. Fig. 5-8) Projections between areas of visual cortex are always reciprocal. Projections to higher

148

levels of visual cortex tend to be very specific and reciprocal projections more diffuse. All V1 projections to secondary visual cortex are a precise retinotopical array and all secondary visual cortex areas are retinotopically mapped.

Each successive neuron in the visual path has ever more complex pattern sensitivities and larger receptive fields. These complex pattern sensitive neurons remain sensitive to line orientation but are less sensitive to particular retinal positions. By the time the visual signal enters V5 in the medial temporal lobe, the retinal field of the neurons usually encompass the entire fovea area.

V2 visual cortex surrounds the primary visual cortex and is made up of a regular array of stripes. Thick stripes are a part of the magnocellular path involved with the detection and discrimination of motion. Thin stripes are a part of the parvocellular path involved in the detection and discrimination of color. Inter stripes are a part of the parvocellular path involved in the resolution of form. These distinct visual pathways are connected at various levels of the visual system but are distinct through the V2 cortex.

V3 visual cortex is completely contained within the sulci of the occipital lobe and is therefore not visible on the exterior surface of the cortex. V3 cortex is a part of the magnocellular pathway involved with motion. V4 visual cortex is located in the second gyrus on the lateral side of the brain and is exposed on the visible surface of the occipital cortex. Neurons within V4 are concerned with color discrimination and many of these neurons exhibit selectivity to line orientation. Lesions of V4 cortex typically result in a loss of color vision.

V5 visual cortex is often referred to as medial temporal cortex (MT). This area of cortex is located in the second sulcus on the lateral side of the brain and is not visible on the exposed surface of the cortex. The neurons of V5 are concerned with motion and the direction of the motion within their retinal field. Patients with lesions to area V5 can experience a complete loss of detection of motion. Objects that move vanish from the conscious visual image of these patients and they live in a static world where objects disappear and then magically reappear.

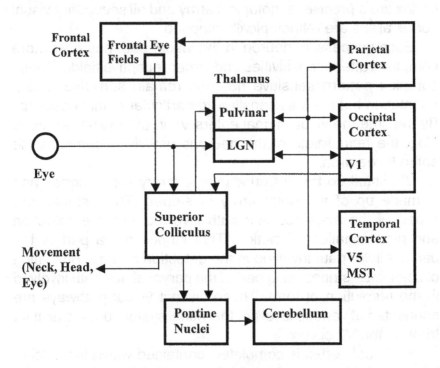

Figure 5-9: Superior Colliculus System

The superior colliculus exists in all mammals, is concerned with spatial orientation, and controls saccades that direct the eyes to a desired location in the environment. It receives input from the retina via the optic nerve, the frontal eye fields of the frontal cortex, and the primary visual cortex. (Ref. Fig. 5-9) Retinal input serves to construct a visual map within the superior colliculus that is combined with audio and somatosensory maps to construct a spatial representation of the immediate environment. This information is projected to secondary visual cortex areas via the pulvinar nucleus of the thalamus. Visually related cortex in the parietal lobe concerned with spatial representation projects strongly to the frontal cortex. The frontal cortex combines this spatial information with a great deal of other input to make conscious decisions about where to direct the eyes. Commands directing eye movements are projected from the frontal eye fields of the frontal cortex to the superior colliculus.

Visually related cortex in the temporal lobe concerned with movement in the visual field projects to the pontine nuclei and thereby to the cerebellum. The cerebellum projects to the oculomotor nucleus in the brain stem to assist in eye control for smoothly tracking moving objects. This supports a form of eye movement that we have not yet discussed called pursuit eye movement. Visual attention to moving objects is obviously important to survival and your visual system has the capability to track motion in a smooth way devoid of jerky saccades. Place your finger approximately two feet in front of your eyes and move it back and forth across your field of vision at various speeds. Notice your eye and head movements associated with this visual smooth tracking. Now remove your finger and attempt to smoothly move your eyes to track an imaginary finger movement. Saccades are the result.

The neural circuitry involved in pursuit eye movements involves direct connection from motion cortex areas V5 and MST of the temporal lobe to the pontine nuclei of the brainstem. The pontine nuclei project to the contralateral cerebellum that modulates the motor projections of the superior colliculus to achieve the smooth eye motion. Without a moving target signal and the corresponding cerebellar modulation, smooth movement of the eyes is very hard to replicate.

This entire system is involved in attention mechanisms. The eyes are directed precisely to whatever in the environment you wish to pay attention. You commonly close your eyes to block out external input when you wish to internalize your attention. Eye saccades to loud and startling inputs are reflexive and saccades to movement are close to reflexive.

I find it fascinating that this ancient hardwired brain sits in the middle of our brains and provides exactly the same function for us it has always provided in exactly the same way because, it is exactly the same.

Many more neurons in the temporal lobes participate in object recognition than the number of neurons in the parietal lobes involved in spatial properties. The inferior temporal cortex receives visual input from secondary visual cortex and is composed of repeated sets of columns. The retinal field of inferior temporal cortex neurons includes the entire retinal fovea and object recognition usually encompasses an eye saccade to the object of attention. The inferior temporal cortex is not retinotopically mapped and specific clusters of neurons identify objects regardless of their retinal location.

How you visually recognize objects has long been a subject of intense research. You can recognize a teapot at any angle of view, from a partial view, a distorted view, a degraded partial image, and also identify the many different shapes that teapots come in. This is an amazing ability far beyond the capability of any man made equipment. Visual object recognition is accomplished by means of straightforward pattern recognition. All of the visual patterns that have ever caused recognition of a teapot have been incorporated into the cerebral pattern that defines a teapot

The key to object recognition is the detection of all edges within the retinal image. The edges of an object are not the only key to recognizing what is an object. Areas of the retinal image that have the same color, the same texture, luminance and same motion are also keys to object identification. Depth perception also helps to identify visual input related to the same object.

The highest levels of object recognition cortex receive input from the temporal and parietal lobes. This is where visual information from object recognition and spatial information come together. This cortex is not topographically organized. Its neurons have very large retinal receptive fields, essentially the entire receptive field. This is the final object identification regardless of where in the retinal image the object lies. Approximately 45% of the neurons here fire in response to the visual image of particular objects, I.E. faces, hands, etc.

The detection of motion is perhaps the first visual capability to evolve. Your eyes are attracted to motion by reflex neural circuits in your brainstem. Primitive vertebrates such as the frog only detect movement and objects that do not move are transparent to them. This same lack of visual capability occurs in the periphery of your retina. You notice movement in retinal fields at the edge of your vision but are unable to identify objects there. Peripheral objects that cease to move drop from your perception. Movement in retinal fields at the very extreme edge of the retina are not perceived consciously but do cause reflex circuits to move the head and eye to bring the moving object into your field of view.

The M type ganglion cells of the retina and magnocellular neurons of the thalamus are not sensitive to motion. The first neurons that detect motion are in V1. Neurons in columns of orientation for each retinal field are sensitive to motion perpendicular to their angle of sensitivity. The output of V1 hypercolumns signal detection of movement perpendicular to all angles of orientation for all retinal fields. This dissected motion information is reassembled in the visual path. In both V5 (MT) and MST there are neurons that are sensitive to both direction and rate of motion. V5 contains a map that signals speed and direction of all moving objects within the retinal view.

The human eye is sensitive to the narrow spectrum of wavelengths from 400 to 700 nm that defines our notion of visible light. Of mammals, only primates are sensitive to divisions within that spectrum that allows them to differentiate the visual field by wavelength or color. The visual systems of non-primate mammals only have one pigmentation in their visual receptors as opposed to humans who have three. Those three pigments differentiate three different types of cones in the human retina. The three pigments cause the cones to be sensitive to different parts of the visible spectrum that define the colors blue, green, and red. We therefore define these colors as primary colors. All of your color discrimination is built from different combinations of these three colors. A cone having a pigment sensitive to red means that the cone is more sensitive to red light and therefore more likely to fire in response to red light than other light.

The evolution of human color vision can be traced by comparing the state of color vision in our primate cousins and by comparing the amino acid sequences of the visual pigments involved. All three-color pigment genes derived from the common cone rhodopsin gene. The first gene was more sensitive to the color blue and the blue gene further differentiated to be sensitive to the color red. This two color system sensitive to blue and red remains the visual state of new world monkeys that divided from old world monkeys around 30 million years ago when the continents of Africa and South America divided. The single long wave gene for red further divided to support two separate pigments sensitive to red and green wavelengths. This three-color system is the current state of both old world monkeys and humans.

The human color vision system works basically by opposition comparisons of the three color cone inputs across the retinal fields. These comparisons are made analogously to the light and dark center surround patterns used in edge detection. Retinal ganglion and thalamic neurons with circular retinal fields are tuned to the color sensitive cones in the dot and surround pattern. On and off center neurons exist for the color comparisons. Many neurons compare on and off center patterns for red and green that produces the sensation of yellow when received equally. Comparisons between yellow and blue form another important comparison discrimination. In the receiving neurons of the blobs of V1 the input pattern sensitivities are more complex. On and off center patterns are detected that combine on and off combinations of two primary colors in both the center and surround pattern. In this manner the blobs of the hyper-columns signal the frequency makeup of the entire retinal image to secondary cortex for further pattern detection.

What is most amazing about your visual system is that you perceive the world as a continuously changing dynamic world that you see in real time. Your visual input comes in the form of static snap shots of this world as a result of eye saccades and synchronous projections from your retinas. Your visual perception of the world around you comes from the neural representation you have built inside your head.

You are not aware that visual input is two-dimensional or that it is a series of snapshots. Your brain creates a three dimensional world and fills in the visual input holes created by saccades. Moving objects continue to move smoothly as you glance around. You perceive the world as you have learned to expect to perceive it.

Auditory Input (How You Hear Your World)

Humans utilize auditory input for three major purposes; language, music, and spatial location of sound sources. The ability to locate the precise angular position of a growl is an ability that has been very highly selected for by evolution. The result is an auditory input system that pattern detects audio timing and intensity differences from both ears to glean this information precisely. Auditory input from both ears is provided to every level of the audio input path. The neural structures from brainstem to cerebral cortex that perform audio spatial localization are the same in all mammals.

Language and music are two types of information that can be transmitted in an auditory form. The receipt of these types of audio signal energy is similar in all mammals. The extraction of the information contained in the auditory signals that comprise music and language is unique to humans. The primary auditory cortex pattern detects fundamental auditory patterns for each input frequency and these auditory patterns facilitate the pattern recognition of language and music by higher-level cortical areas.

Auditory input from the organ of corti projects to the cochlear nuclei in the medulla oblongata (Ref. Fig. 5-10). The cochlear nuclei lie external to the inferior cerebellar peduncle in the medulla oblongata and consist of three subdivisions. Each of the 30,000 inputs from the organ of corti carries neural input reflecting the detection of one particular frequency. All input from the organ of corti concerning the full frequency spectrum project collaterals to all three nuclei.

Note the number of nuclei in figure 5-10 that pattern detect input from both ears. The trapezoid body lies in the middle of the medulla oblongata and integrates auditory input from both ears. The superior olivary nuclei are key to the localization of sound input. The lateral portion of the superior olivary nucleus is concerned with differences in the intensity of sound energy between the two ears and the medial portion of the superior olivary nucleus is concerned with the differences in timing.

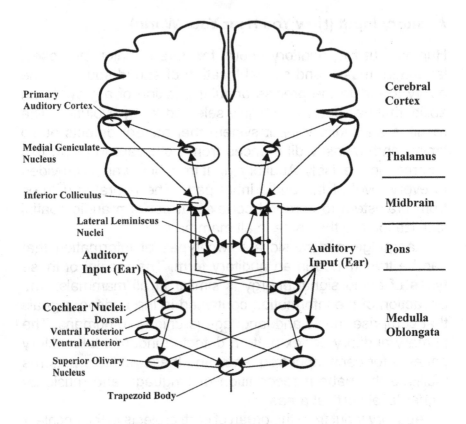

Figure 5-10: Auditory Input

At the level of the pons the lateral leminiscus nuclei receive collaterals from inputs bound for the inferior colliculus and project strongly between themselves. Lateral leminiscus projections terminate in the inferior colliculus of the midbrain. As you remember from our visual system description, the colliculus serves as a spatial map integrating visual, somatic and auditory inputs in order to control eye movements. A loud noise causes the colliculus to direct the eyes to the source of the noise in a reflex action. The localization of sound is delivered to the colliculus in a spatial audio map array. The inferior colliculus projects to the ipsilateral medial geniculate nucleus of the thalamus (Ref. Fig

5-1) that performs its normal switchboard function. The medial geniculate nucleus projects to the primary auditory cortex in the superior temporal gyrus. (Ref. Fig 5-11)

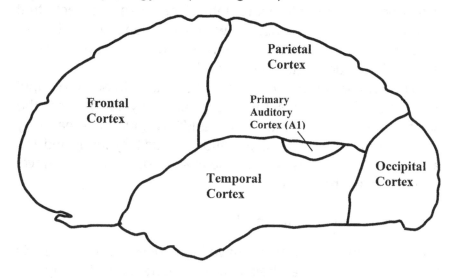

Figure 5-11: Primary Auditory Cortex

The primary auditory cortex has a columnar organization with summation columns receiving input from both ears and suppression columns dominated by one ear. Contralateral callosal audio input is received in callosal input zones alternating with zones of no callosal input much like ocular dominance columns.

It is interesting to view Figure 5-10 as two separate systems. The old auditory system from the midbrain down localizes the input of sound based on timing and intensity differences between the two ears. This information is built into an auditory spatial map in the inferior colliculus. This old system is a complete auditory system including the feedback necessary for attention and control. The new auditory system is above the midbrain and regulates the old system by feedback to the inferior colliculus. The new auditory system of thalamic and auditory cortex sits on top of the old brain stem auditory capability.

Auditory cortex projections to the parietal cortex support the locating of sounds in your environment. You can also identify objects by how they sound. A dog's bark or a cat's meow is as effective in identifying these animals as seeing them. As in visual object recognition, audio object recognition is most developed in the left temporal lobe. One class of objects that humans learn to recognize is words and language. The left temporal lobe is where you build the object recognition patterns that enable your language capabilities.

The primary auditory cortex pattern detects auditory input into discrete primitives for each frequency range and the structure and output of the primary auditory cortex is remarkably similar in all humans. Interpretation of auditory input and the acquisition of language and music abilities occur in higher-level cortical areas through learning.

Chemical Input (Taste and Smell)

Your chemical input systems have been inherited from animals that first evolved over half a billion years ago. The age of these chemical input systems is reflected in their uniqueness. Olfactory input is the only input type that synapses directly to the cerebral cortex without first synapsing with the thalamus. Gustatory input does synapse in the thalamus prior to the cortex but its major input does not cross the midline and enters the ipsilateral thalamus and cortex. Neither of these systems exhibits any linearity with respect to the chemicals they detect.

The gustatory input path begins with taste buds innervated by receptor ganglion neurons that synapse in the gustatory nucleus in the medulla oblongata. The axons of the gustatory nucleus ascend ipsilaterally to the ventral posterior medial nucleus of the thalamus that also deals with somatosensory input from the contralateral tongue area. The taste input path is closely aligned with the somatosomatic input from the tongue and mouth in both the thalamus and cerebral cortex.

Olfactory input begins with chemical receptor neurons contained in patches located in the lining of the nose. The axons of these neurons synapse in the olfactory bulbs just below the frontal cortex. This olfactory input synapses on structures within the olfactory bulbs called glomeruli. The number of neurons contained within the glomeruli is much less than the number of receptor neurons in the nose.

The olfactory bulb contains three distinct regions; medial, lateral, and intermediate. All regions of the olfactory bulbs project to old cortical areas and from there to other brain structures. The medial and lateral regions provide input to the limbic system and the intermediate region provides input to the thalamus. The lateral region projects to old cerebral cortex in the temporal lobe that acts as the primary olfactory cortex. The medial region forms reciprocal connections with the frontal lobe.

Balance

The final neural input type utilizes input from the inner ear indicating head movement and position to coordinate balance. The vestibular nuclei of the medulla oblongata receive this input and project to the cerebellum. The cerebellum produces learned output that maintains posture and balance during movement.

All of your sensory inputs are utilized to construct a neural representation of your world inside your head. That virtual world is three-dimensional and contains a variety of objects that move in a smooth continuous fashion. You have learned what objects your world is made of and you have learned how those objects behave. Much of your neural representation of the world comes from what you know and not from what you actually perceive.

Component #1 – The Spinal Cord
(Your Body's Interface to Your Brain)

The functions of the spinal cord are the same over its entire length. Physical differences in spinal cord segments are due to the neural requirements supported and the amount of axons transversing through that segment. The spinal cord's gray matter is partitioned according to function into ten areas called laminae. (Ref. Fig. 5- 12) Laminae one through six receive input, laminae seven and eight contain interneurons concerned with modulation of input and output, laminae nine houses motor neurons and laminae ten is the spinal cord's gray commissure.

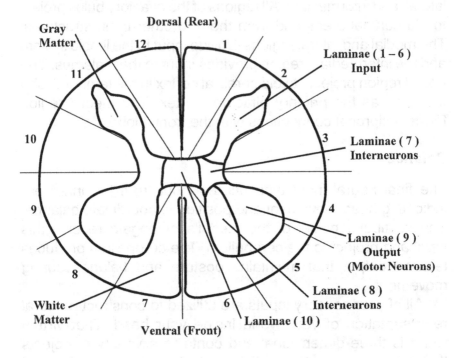

Figure 5-12: Spinal Cord – Laminae

The spinal cord develops from the lower two thirds of the neural tube in the growing human embryo. The alar plate matures to become the entire rear input side of the spinal cord and the basal plate becomes the frontal output side. The evolution of the spinal cord began in the Cambrian period with Chordates over 500 million years ago. Spinal cord functions of primitive pain avoidance, reflex actions, coordination of opposing muscle group output, and support of rhythmic movements are implemented by spinal cord interneurons and are the same in all vertebrates.

The spinal cord supports all somatosomatic input from the entire body. Proprioceptive and sensory input enters the dorsal side of the spinal cord and either ascends to the brain stem or synapses with secondary neurons in spinal cord laminae one through six. Many of these sensory input axons and the axons of neurons in laminae one through six project collaterals to the interneurons of the spinal cord in laminae seven and eight.

The corticospinal tract contains the cerebral projection emanating from pyramidal neurons in the motor and somatosensory cortices. Approximately one half of these outputs terminate on the motor neurons in laminae nine and support voluntary muscle control. The other half provides inhibitory feedback control of somatosensory input. Interneurons in the spinal cord pattern detect output projections from brainstem nuclei and project to motor neurons in laminae nine to control muscle activity.

The spinal cord contains an impressive amount of neural material and performs a wide range of functions concerning neural input and output. Lets quickly examine three of these spinal cord functions to get a feel for the kind of neural processing performed by the spinal cord. The three functions we will cover are pain avoidance, the knee jerk reflex and opposing muscle inhibition.

Pain avoidance is a primitive reflex action implemented by all segments of the spinal cord. When your hand comes into contact with something hot or sharp that causes pain, that pain signal propagates into a rear horn of the spinal cord and provides collateral synapses with interneurons. The axons of these interneurons signal directly to motor neurons in laminae nine that fire and cause the hand to be retracted. By the time the pain signal is acknowledged by the brain, the muscles controlling the location of the hand have been activated in this primitive reflex action.

When a doctor wants to verify the integrity of your spinal cord, he or she commonly performs the knee jerk reflex test. The leg is relaxed in a sitting position with the foot hanging freely. With a small hammer, the doctor taps the large ligament just below the kneecap connecting the shinbone with the thigh muscle. If everything is working properly, the lower leg responds by jerking forward in response to the tap.

When the hammer strikes the ligament it activates proprioceptive sensors in the ligament that sense elongation of the tendon. This tendon elongation signal is propagated to a rear horn of the spinal cord and is received by interneurons. Spinal cord interneuron circuits exist that maintain stability and muscle tension to avoid falling. The tendon elongation signal causes the interneurons to signal to the motor neurons that project to the thigh muscle to contract and counter this sudden elongation collapse of the lower leg. This reflex signal to the thigh creates the observed knee jerk. In this manner the doctor is able to quickly test the integrity of proprioceptive input, interneuron response and motor output of your spinal cord.

Interneurons in the spinal cord provide opposing muscle inhibition that allows freedom of movement. Spinal cord interneurons provide this basic function by inhibiting the motor neurons of opposing muscle groups when a particular muscle group is activated. The activation of motor neurons and muscle fibers always causes the inhibition of motor neurons controlling opposing muscle fibers. This inhibition is an automatic spinal cord function implemented on all the motor neurons controlling the muscles of the body. Without it, coordinated movement would be impossible.

Component #2 – The Human Brain Stem (Your Old Brain)

The one component that literally enables you to be alive is the brain stem. Your brain stem controls your unconscious neural systems and a properly functioning brain stem is absolutely required for you to exist. Despite its required functionality, your brain stem is as hardwired as a crocodiles. Accordingly, we will cover this second pass of your brain stem component as efficiently as possible.

We will cover each of the brain stem areas, the medulla oblongata, the pons, and the midbrain, examining schematics of each area in order to see the major nuclei that populate the area and the major neural tracts that transverse the area. Lastly, we will also cover the reticular formation that exists in each of the brain stem areas and is pervasive in controlling overall brain function.

The medulla oblongata closely resembles the spinal cord in both development and function. The alar plate in the embryo develops into the cranial nerves and nuclei and the basal plate into the motor neurons that output to the muscles of the head. Most of the basic functions necessary for life to exist in a complicated life form are implemented by the medulla oblongata as we might expect from the first primitive brain.

In the following discussion of the various structures and related functions within the medulla, we will follow closely the schematics of Figures 5-13 and 5-14 that contain the various medulla nuclei and neural tracts that transverse the medulla

There are five distinct inputs from the body. The most primitive, pain and temperature, enters the medulla and moves to a more rear lateral position as it passes through the medulla. Pain and temperature input from the head enters the medulla, synapses in the spinal trigeminal nucleus and ascends to the thalamus. Axons carrying information about pain and temperature give off many collaterals to the reticular formation.

A more modern input supporting more precise location of pain and temperature and poorly localized touch from hairless areas of the body enters the medulla bound for the somatosensory nuclei of the thalamus. There is extensive branching of collaterals from this tract to the reticular formation. There is so much branching of this input type in the medulla that only approximately one third of the ascending axons entering the medulla actually reach the thalamus.

Your most recently evolved somatosomatic input, well-localized touch, movement and position, enters the medulla in the rear columns. These sensory neuron axons first synapse in nuclei in the lower medulla oblongata. The axons of these nuclei cross the midline and ascend to the thalamus in the lateral portion of the brain stem. Proprioceptive input enters the medulla bound for the inferior and superior cerebellar peduncles.

Descending motor output through and from the medulla oblongata is all destined for the motor neurons or interneurons of the spinal cord. The massive tract of axons from the pyramidal neurons of the motor cortex enters the upper medulla in the

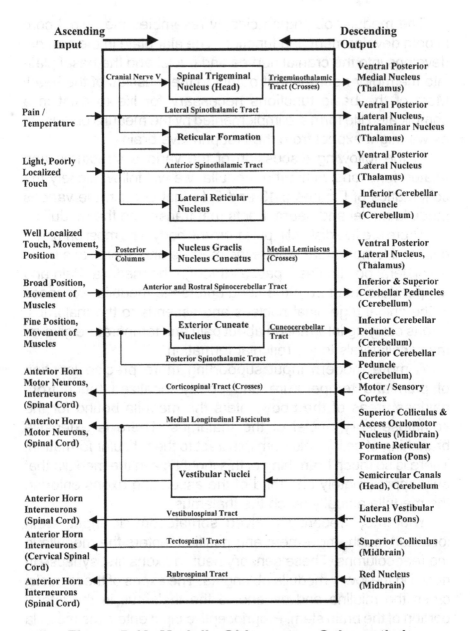

Figure 5-13: Medulla Oblongata – Schematic I

most forward position. This tract proceeds down to the lower medulla where it then travels posteriorly and laterally to proceed down to the spinal cord. Approximately 85% of these fibers cross the midline at this lower level of the medulla.

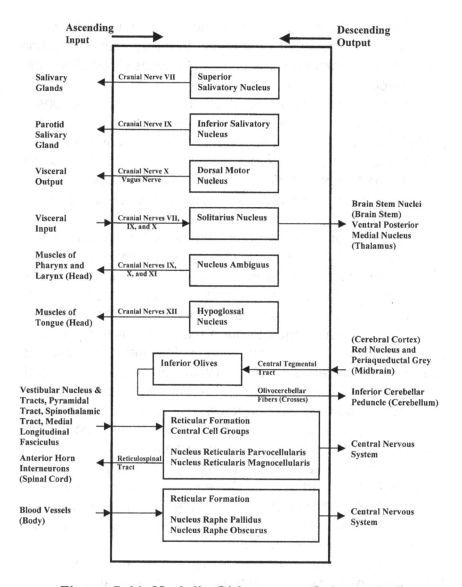

Figure 5-14: Medulla Oblongata – Schematic II

There are four vestibular nuclei in the medulla oblongata that are concerned with balance and they all receive their input from the semicircular canals of the inner ear. These nuclei are located from the mid medulla to the mid pons and control your equilibrium. The vestibular nuclei have extensive reciprocal connections with the cerebellum. The vestibular nuclei project to many areas of the brain stem and also to the thalamus. Vestibular projections to the reticular formation cause motion sickness during times of excess stimulation from the semicircular canals.

The medulla oblongata regulates the autonomic functions of the body and contains a number of neural control centers that perform this function. The superior and inferior salivatory nuclei lie near the midline in the upper medulla and control the salivary glands. The dorsal motor nucleus and solitarius nucleus are the output and input portions of many of the feedback control loops that regulate the body's visceral functions.

Dorsal motor nucleus output fibers exit the medulla and proceed down the neck to the chest and abdomen to synapse on post ganglion neurons in the walls of the various organs. The specialized muscle tissue of the heart is innervated and a slowing of the heart rate is effected. The entire digestive tract from the esophagus through the gastrointestinal tract is innervated. The rhythmic contractions of the intestinal walls and the secretion of glands associated with digestion are controlled. The smooth muscles of the lungs are innervated controlling the process of breathing. The pancreas, gall bladder and liver are also on the extensive list of organs under the influence of this group of neurons in the medulla oblongata.

The solitarius nucleus is the input portion of the feedback control loops utilized to control breathing, digestion and heart function. Sensations of taste are received from the tongue and pharynx. The solitarius nucleus projects to many nuclei in the brain stem and also to the thalamus.

The nucleus ambiguus contains motor neurons that project to the pharynx and larynx controlling the act of swallowing. This motor output also controls the vocal cords. The hypoglossal nucleus contains motor neurons that control the muscles of the tongue. Both the ambiguus and hypoglossal nuclei are instrumental in the production of speech.

A major component of the motor output control system of the CNS is the inferior olives of the upper medulla. The inferior olives receive collateral inputs from motor output and projections of the inferior olives cross the midline to enter the inferior cerebellar peduncle as climbing fibers. These inputs from the inferior olives make up the largest portion of input in the inferior cerebellar peduncle.

The medulla oblongata provides an analogous function as the spinal cord for somatosomatic and proprioceptive input. For motor output it provides a path for upper brain motor centers, posture control via the vestibular nuclei and an important way station in the motor output feedback path via the inferior olives. It shares with the rest of the brain stem the all-important reticular formation. Finally, it provides the control of heart, breathing, digestion, blood pressure and all of the other unconscious functions necessary for life in a multicelled animal.

The rear portion of the pons closely resembles the medulla in both structure and function. The forward portion carries the large tracts of axons bound for the middle peduncle of the cerebellum. As in the medulla, we will closely follow the schematics of figures 5-15 and 5-16 in the following discussion. The inferior cerebellar peduncles exist both in the upper medulla and in the lower pons. With the exception of the input signals lost to the inferior peduncle and nuclei in the lower medulla, the input portion of the medulla and pons are identical. The motor output portion of the pons is also very similar to the medulla. At this level the corticospinal tract spreads out to pass through the pontine nuclei giving off collaterals to them as it passes. The axons of the pontine nuclei cross the midline to enter the cerebellum as the middle cerebellar peduncle.

Facial input concerning pain and temperature and some somatosensory information enter the pons via cranial nerve V to terminate in the spinal trigiminal nucleus. Axons from the spinal trigiminal nucleus cross the midline and ascend to the thalamus. Facial input concerning pressure and two-point discrimination also enter via cranial nerve V and terminate in the main sensory nucleus. The axons of this nucleus also ascend to the thalamus as crossed and uncrossed fibers. Proprioceptive input concerning chewing is input from the muscles of the jaw to the mesencephalic nucleus. The motor nucleus of cranial

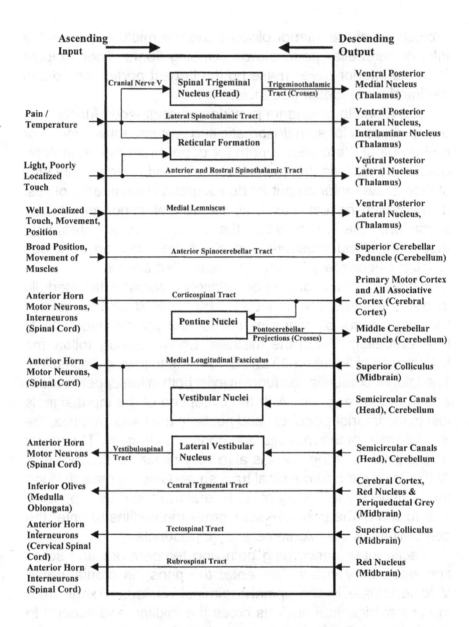

Figure 5-15: Pons – Schematic I

nerve V exerts output control over the inner ear, the pharynx, the muscles that close the mouth, and also plays a role in chewing and swallowing. Neural reflex control associated with the fifth cranial nerve includes sneezing, vomiting, blinking of the eye, and tearing of the eye.

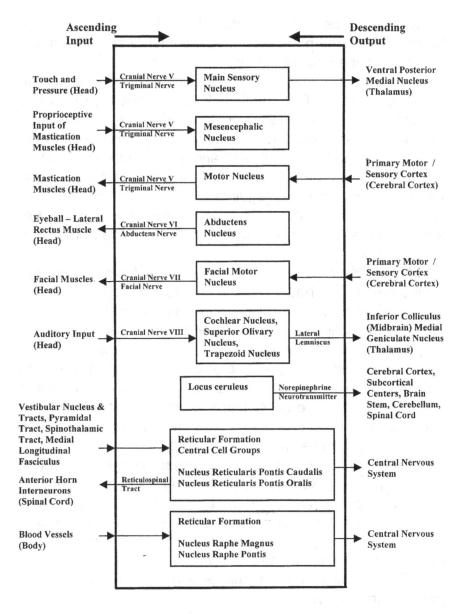

Figure 5-16: Pons – Schematic II

Cranial nerve VI carries output control from the abductens nuclei to a muscle controlling the eyeball. The facial motor nucleus receives voluntary control input from the motor cortex and drives the muscles of facial expression. The auditory input from the ears enters via cranial nerve VIII and synapses on nuclei in the pons and upper medulla. Output projections from these nuclei in the pons ascend to the inferior colliculus in the midbrain concerned with the location of sound.

The locus ceruleus nucleus is located in the rear portion of the upper pons. This small nucleus provides approximately 50% of the neurotransmitter norepinephrine to the entire CNS. The axons of the locus ceruleus travel large distances to the cerebral cortex, the spinal cord and everywhere in between. Norepinephrine is involved in the excitation of neurons and the locus ceruleus helps regulate your level of alertness. This small neural center is also involved in the circadian cycle of sleep and rest and plays a key role in determining to what you pay attention.

The midbrain provides the higher-level functions of your reptilian brain. The housekeeping chores provided for the body are mostly provided by the medulla and pons. The midbrain contains the colliculus nuclei where the first representation of the outside environment is built. The colliculus projects directly to motor neurons to control motor output in response to loud noises and visual stimuli. This spatial capability is hardwired and there is no evidence of learning within the midbrain. The schematics of figures 5-17 and 5-18 should be referenced for the following discussion.

Ascending somatosomatic input passes through the midbrain on its way to the thalamus. Collateral input carrying information concerning pain synapses in the periaqueductal gray nucleus in the rear medial area of the midbrain. The periaqueductal gray nucleus receives input from the cerebral cortex, pain feedback input from the thalamus and is involved in the suppression of pain. This nucleus signals to the reticular formation to affect its pain suppression role.

There are three cranial nuclei located in the midbrain, the oculomotor nucleus, the trochlear nucleus and the mesencephlic nucleus. The first two are output nuclei involved in the control of the eyeball and the third is an input nuclei receiving proprioceptive input from the head.

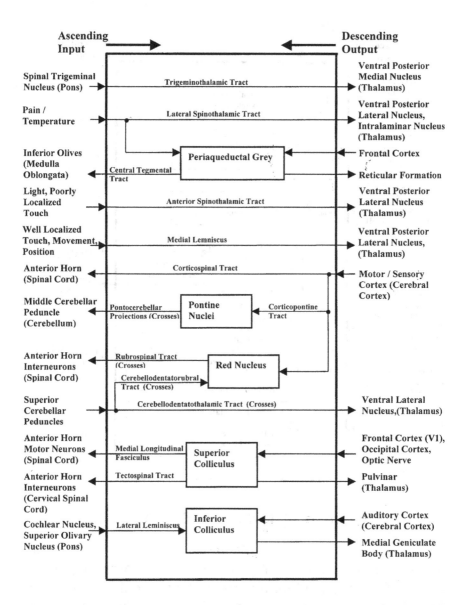

Figure 5-17: Midbrain – Schematic I

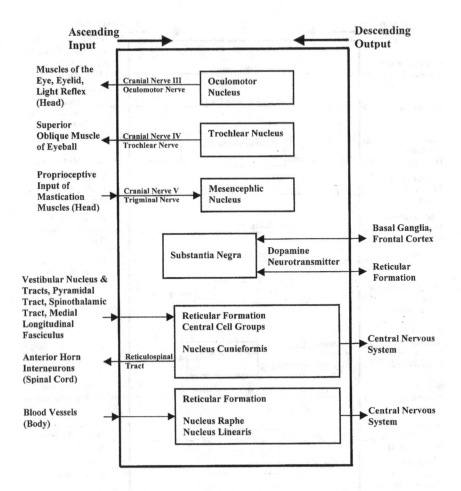

Figure 5-18: Midbrain – Schematic II

The Substantia Negra is the largest nucleus in the midbrain and is the major producer of the neurotransmitter dopamine. Dopamine is required for the learning of output behavior. The frontal cortex and the basal ganglia are the two main motor output areas that interact with the substantia negra in the control of dopamine output.

The reticular formation regulates your alertness and activity level enabling you to cope with the current state of the environment. The reticular formation acts as a central control center that determines what input stimulus requires attention and possibly action. Pain is only the most obvious of the types of input stimulus that causes the reticular formation to exert control.

This complex system is made up of three major groups of neurons with at least thirteen separate nuclei among them. These various nuclei can be seen in Figures 5-14, 5-16, and 5-18.

The seven nuclei that make up the raphe complex are located near the midline of the entire brainstem. These nuclei receive input from small blood vessels. Their axons produce serotonin, cover great distances and influence the entire neural system. It seems that these neurons are monitoring the chemical content of the blood in order to affect their neurotransmitter control.

There are five nuclei that make up the central reticular complex. These nuclei are the largest of the reticular formation and contain neurons with extensive dendritic trees. These neurons receive input from the various neural tracts that pass through the brain stem and signal widely to the entire nervous system. The lateral reticular nuclei exist in the lower brain stem and receive pain signals from the spinothalamic tract in order to affect the level of alertness that pain usually causes. The lateral reticular nucleus projects to the cerebellum through the inferior cerebellar peduncle. The reticular formation forms neural circuits with much of the brain and performs an enabling function that is required for normal operation.

The brain stem represents an entire brain from one point in evolution. Many of the functions of this primitive brain are still intact and unchanged in the human brain. With continued evolution some of the functions of the brain stem have come under the conscious control of higher-level brain systems. The things that are unchanged are the things you don't have to worry about as long as the brain stem does.

Output control of your muscles is a different story. The brain stem output control areas such as the colliculus are an integral part of your skeletal neural control system. These output systems are built level upon level with each level exercising some but not complete control over the lower levels.

The brain stem is in itself a hierarchy of neural control. Most of the housekeeping functions of the body are handled by the medulla oblongata in the lower brain stem. As we progress further up the brain stem we see fewer of these housekeeping functions. Due to its location, the pons performs the role of the medulla for the necessary functions of the head. The pons plays a key role in the output control system through its pontine nuclei and providing the interface to the cerebellum. The midbrain builds a representation of the external environment and directs actions to deal with it.

This primitive brain contains rudimentary neural control that is still utilized by the highly evolved human brain. The locus ceruleus, substantia negra and reticular formation all perform functions that control the level of neural activity in the entire brain stem as well as the rest of the CNS. These integrated structures react to pain, sexual stimulation, hunger, or thirst and stimulate the neural system to do something relevant to the situation. These systems are intact and unaltered by evolution. The things that motivate reptiles still motivate you.

Component #3 - The Cerebellum
(Motor Output Memory)

I am going to assert that your cerebellum is the actual driver of most of what we have termed voluntary movement. Your cerebellum learns all of the pattern states of the cerebral cortex that elicit skeletal movement and stores the motor output patterns associated with those states. When the cerebellum recognizes a cerebral pattern that will cause skeletal muscle output, it projects the learned motor output pattern to the motor cortex, M1, through the thalamus. The learned motor pattern from the cerebellum is used by M1 to directly drive motor neurons unless you consciously override it. Your cerebellum is your autopilot. Lets examine the cerebellum and why I make this assertion.

The cerebellum does not project directly to the motor neurons that drive muscle cells. It modulates those motor neurons by providing input to the neural structures that produce skeletal motor output. The cerebellum supports three areas of motor control. The oldest area, the archicerebellum, deals with balance and posture. The paleocerebellum interacts with brainstem output nuclei that drive non-voluntary motor output. The newest area, the neocerebellum, evolved in conjunction with the cerebral cortex and interacts with it in the control of voluntary movement.

The cerebellum is a very old brain structure that exists in fish and is well developed in reptiles. Its neural circuitry and function are the same in all vertebrates. The cerebellum appears to be the first brain area to evolve learning capability by developing the ability to modify its pattern detection to recognize neural patterns generated through experience.

The cerebellum physically consists of an outer cortex sheet surrounding white matter and cerebellar nuclei. The cerebellar cortex contains three layers of neurons with a middle layer of large purkinje neurons that are the output neurons of the cortex. The outer layer of the cerebellar cortex is an orthogonal matrix formed by the horizontal axonal projections of granule neurons located in the inner granule layer and the vertical dendritic arbors of the purkinje neurons. This regular structure allows each purkinje neuron's dendritic arbor to receive projections from around 200,000 granule neurons.

There are approximately ten billion granule neurons in the inner granule layer of the cerebellum. This vast number of granule neurons causes the cerebellum to contain over 50% of all the neurons that make up the human brain. Ten billion granule neurons packed into the limited space of the cerebellum gives you an idea of the small size of the granule neuron. The cerebellum contains proprioceptive body maps over its surface area, reminiscent of the cerebral somatosomatic input area (S1). These maps are not a linear representation of the body as in the cerebral cortex but are a broken and jumbled representation of the body.

In the developing embryo, the cerebellum grows from the dorsal alar plate consistent with all input areas. The cerebellar cortex forms and then expands horizontally like a balloon to fill all of the available space. Like the cerebral cortex, this balloon continues to expand after it runs out of space and folds into gyri and sulci. The regular horizontal granule neuron axonal structure causes the fine folds of the cerebellum to take on their horizontal nature.

There is continuous output from all of your motor output systems to the skeletal muscles of your body. Every one of your motor neurons receives input directly or indirectly from every neural structure that provides skeletal motor output. The individual neural circuits driving each motor neuron are serviced by a particular set of cerebellar circuits. Those cerebellar circuits provide learned output for balance, non-voluntary skeletal control and voluntary movement that modulates the output neuron's firing.

The neural circuit that comprises the entire cerebellum is depicted in figure 5-19. The purkinje layer is filled with continuous rows of purkinje neurons that make up a sheet of neurons. Purkinje neurons receive four different kinds of inputs. Climbing fibers exert strong excitatory input on both the cell body and the dendritic arbor. A firing climbing fiber always causes the purkinje neuron to fire. Basket cells exert strong inhibitory input on the cell body of the purkinje neuron. Basket cells sense the firing of purkinje neurons in their dendritic vicinity and inhibit neighboring purkinje cell bodies. Stellate neurons in the molecular layer also exert inhibitory influence on the dendrites of purkinje cells. This form of neural competition has the effect of one purkinje pattern at a time gaining dominance over all others.

The final input to the purkinje neuron is the axonal projections from granule cells to the purkinje cell's dendritic arbor. Granule axonal projections run horizontally through the molecular layer of the cerebellar cortex in extremely large numbers, approximately ten billion. The dendritic arbors of the purkinje neurons resemble the branch of a tree that grows only in the vertical plain to resemble a fan. The large numbers of horizontal granule neuron axonal projections run through all areas of this fan in a perpendicular fashion and allow the purkinje neuron to receive an extremely large input from granule neurons. These synapses terminate on the dendritic spines of purkinje neurons. Dendritic spines also exist on pyramidal neurons in the cerebral cortex and seem to be the major site for alteration of synapse efficiency in the process of learning.

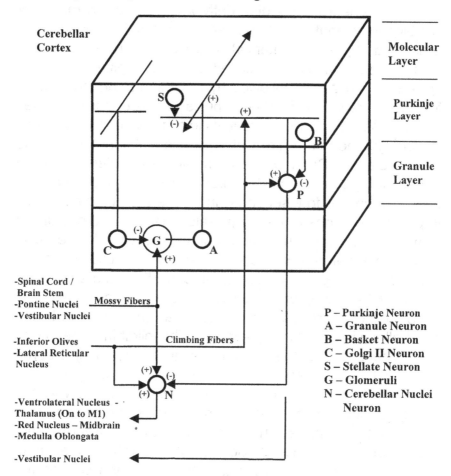

Figure 5-19: Cerebellar Circuitry

Granule neurons are extremely small with short, localized dendritic arbors that receive input from many mossy fibers. Granule neurons pattern detect input from mossy fibers in structures called glomeruli. The actual function of the glomeruli is unknown. They are built with glial cells and glial cells provide much of the structural makeup of the brain. They appear to facilitate the efficiency of mossy fiber and granule cell chemical communication. Inhibitory feedback from Golgi II cells modulates the granule neuron's firing.

Purkinje neurons fire differently when excited by mossy or climbing fiber input. Climbing fiber input is highly excitatory and a climbing fiber action potential always elicits a purkinje neuron to fire with a train of pulses called a complex spike. A complex spike is a large pulse followed by very high frequency pulses for some duration. Mossy fiber input is just the opposite. It takes many mossy fiber action potentials to align temporally to cause the purkinje neuron to fire with a simple spike that is a single action potential. Cerebellar neurons are rarely silent. Climbing fiber projecting neurons in the inferior olives fire spontaneously every second or few seconds and generate complex spikes in purkinje neurons at the same rate. Granule neurons fire at a much higher rate and cause purkinje neurons to fire 50 to 100 simple spikes per second. The normal action potential condition of purkinje neurons is around 75 simple spikes per second with a complex spike every second or so.

Climbing fibers appear to act as an input that puts the cerebellar circuit into learn mode. The climbing fiber firing always causes the purkinje neuron to fire. The purkinje neuron firing inhibits the cerebellar nuclei output neuron from firing. This stops the cerebellar circuit from providing output while the cerebellar circuit is learning. The purkinje neuron firing causes its dendritic synapses that are active to be strengthened. Remember our Mr. Hebb. So an active climbing fiber input causes the cerebellar circuit to stop output and causes the purkinje neuron to learn the current dendritic pattern.

Neocerebellum projections to the red nucleus and on to the inferior olives become climbing fiber inputs back to the purkinje neurons that form the cerebellar learning circuits. The strength of this learning feedback loop is inversely proportional to the strength of the learning for that output procedure. As an event is encountered more often, it becomes more easily and strongly recognized.

Climbing fibers are spontaneously active every second or few seconds. This causes each cerebellar circuit to intermittently stop outputting and transition into learn mode. In learn mode, the cerebellar circuit adjusts its dendritic learning to include the current skeletal output pattern. The cerebellum is altering its learned motor output to incorporate the current motor output pattern from M1. This allows the cerebellar circuit to continually adjust to a body that is growing, aging, gaining or losing weight, and getting stronger or weaker with exercise. Getting out of a chair is a much different movement for a seventy-year old person than for a twenty-year old person. In fact, the movement required to get out of a chair changes continuously your whole life, as does every other movement you make. Your cerebellum is continuously learning the new voluntary output motor patterns associated with these changes.

The cerebellum contains three distinct areas that project to different brain structures. (Ref. Fig. 5-20) These areas perform the exact same neural function. When you raise an arm, all three of the cerebellar areas receive appropriate body input and provide the learned feedback to make the movement smooth and in balance. The archicerebellum is physically located on the lower portion of the cerebellum adjacent to the upper medulla. The paleocerebellum is physically the anterior lobe of the cerebellum located on the top portion of the cerebellum adjacent to the midbrain. The large neocerebellum lobe is called the posterior lobe and occupies the entire middle area adjacent to the pons. The entire cerebellum receives serotonin from the reticular formation to the granule and molecular layers and norepinephrine from the locus ceruleus to all layers effecting alertness and overall neural state.

The archicerebellum interacts directly with the vestibular nuclei. Input from both sides of the body enters both the left and right hemispheres of the archicerebellum. This full body input is required for the maintenance of posture and balance. Counter balancing movement on one side of the body with offsetting movement on the other side is a mainstay of the archicerebellum's operation.

The paleocerebellum receives proprioceptive mossy input from the entire body. This is a massive amount of input containing information from every muscle and tendon in the body and head. Both detailed and broad input about the position and motion of every part of the body are input to the paleocerebellum via the inferior and superior peduncles. The paleocerebellum's purkinje neurons project to the cerebellar nuclei that project to motor output brain stem nuclei.

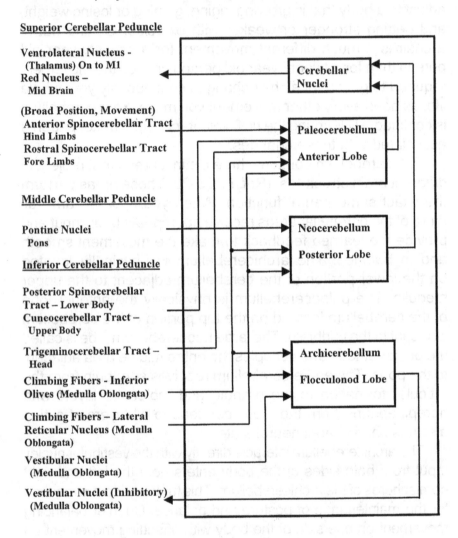

Figure 5-20: Cerebellar Input/Output

The neocerebellum receives a massive amount of mossy input indirectly from the cerebral cortex via the pontine nuclei. Literally, with the exception of primary sensory cortex, the entire cerebral cortex sends projections to the pontine nuclei bound for the cerebellum. The neocerebellum's purkinje neurons project to the cerebellar nucleus that outputs to the premotor and primary motor cortex via the ventral lateral nucleus of the thalamus.

Much of the neural circuitry, structure, and neural interconnections of the cerebellum are well known. Many have postulated that the neocerebellum functions as storage for learned output procedures. However, like most of the brain's components, how the neocerebellum actually works and the complete function it performs is unknown. The information we do have does allow us to apply analysis to glean what the neocerebellum appears to be accomplishing. Lets analyze the neocerebellum's role in voluntary movement.

Clues as to the neocerebellum's function come from the make up of its input and output. The cerebellum receives the state of the entire cerebral cortex. This once pattern detected input includes collateral input of projections from M1 to the motor neurons as well as the overall cerebral cortex pattern associated with that output. The neocerebellum receives both the actual output from M1 and the entire state of the cerebral cortex that caused that output pattern to be selected. The neocerebellum pattern detects that input and projects to the motor cortex through the thalamus.

An important issue is how the primary motor cortex utilizes this cerebellar input. Is the input from the cerebellum used directly as the output or does it modulate M1's output? The attributes of your skeletal muscle control system favor the idea that the cerebellar output to M1 is used directly as the M1 output pattern. The neocerebellum is receiving cerebral output that is driving muscular contractions in real time. Provide feedback to correct output in advance of that output being received is difficult. It is much simpler to just store in cerebellar memory the actual output patterns and provide them to M1 as learned output. The purkinje neurons, under control of climbing fibers, learn the actual M1 output. The cerebellum simply pattern detects the state of the entire cerebral cortex and when it detects planning for motor activity, provides the planned learned motor output back to the motor cortex.

The cerebellum appears to act like a record and playback device. It is recording voluntary muscle control output from M1 and the associated state of the entire cerebral cortex. As an unlearned movement is learned, the cerebellum transitions from record to play. The strength of the play output is directly proportional to the strength of the input pattern detection associated with the movement.

Your conscious control of your muscles is responsive to immediate environmental changes. This fast acting system is driven from procedural memory in the frontal cortex. There is another slow learning system of muscle control provided by the cerebellum. Learned movements can be allowed to drive muscular output directly by passively not asserting conscious control. This system is slow to learn and slow to modify.

What the cerebellum learns is both the cerebral conditions that elicit movement and the motor cortex execution of those movements. The cerebral output associated with planning a movement is a large series of pattern inputs to the cerebellum. Once the cerebellum learns a movement, the recognition of the movement planning will cause the cerebellum to actually send the learned motor output to the motor cortex. This allows the movement execution without conscious intervention. This is what the human cerebellum accomplishes for you.

When you learn a new motor task, such as swinging a tennis racket, every repetitive swing is a completely conscious act until it is learned. As the swing is learned, its control becomes more and more automatic until you no longer have to concentrate on it. If you visit a tennis coach and want to change your swing, you make the movement conscious again and change it repeatedly until the cerebellum learns the new movement.

It is interesting that very little proprioceptive body input reaches the cerebral cortex directly. A system designed to control movement would normally require feedback concerning actual position and movement. Proprioceptive feedback to the cerebral cortex is not required because the cerebellum came before the cerebral cortex and was already performing the task. For mammals, the cerebellum provides learned voluntary skeletal output to M1 when the overall cortical pattern indicates that it is needed. It provides exactly the same function for humans.

The additional capability provided by the human cerebellum is the ability to learn the large and complicated set of human behaviors that include skeletal movement. The complexity and quantity of behavior enabled by the volume of procedural memory space in the human frontal cortex has lead to the ten billion granule neurons required to perform that function. Over half of the neurons in your brain are dedicated to watching what the other half of your neurons are doing in order to provide learned motor output.

Learning to play the piano is a prime example of the human cerebellum at work. You want to learn to play the cord C. This task involves all of the muscles of the upper body. The fingers must be held at a precise arrangement, the arm must be positioned properly and the rest of the body reacts to the movement ever so slightly to remain in balance. The first try is a disaster. The next several tries are not much better but with practice, a couple of hundred, you can play the cord C with relative ease.

Lets fast-forward one year of practice. You now can read music and play the piano with some level of skill. When the sheet music calls for the cord C to be played, you play it automatically without consciously thinking about your fingers. This is also true of all the cords you have learned. If you had to think about how to play the piano, you would not be able to play the piano.

At some point in time you decide to execute a learned task such as hit a golf ball. The cerebral pattern that signals hit the ball is recognized by the neocerebellum. The entire cerebellum is primed to hit the golf ball. These hit the golf ball patterns of purkinje neurons will maintain dominance until the task is completed. A series of neocerebellar projections to the motor cortex through the thalamus drive the actual motor cortex output that accomplishes the swing. As the swing is accomplished, the paleocerebellum and the archicerebellum provide their output projections also learned from practice. This is a learned vanilla swing. If there is an obstacle in the way or some other reason to alter this vanilla swing, the cerebral cortex will consciously assert its dominance over the motor cortex and alter the swing.

The cerebellum performs as a complete neural system. The neocerebellum is pattern detecting two portions of cortical memories during learning. Cerebral patterns that indicate the planning of movement and the cerebral motor neuron output associated with that planning. Once a movement is learned, the neocerebellum plays back the motor neuron output associated with the recognized planning. The paleocerebellum stores the pattern of proprioceptive input associated with that movement and provides learned output for the non-voluntary aspects of the movement. The archicerebellum also stores and recognizes its portion of balance responsibility and signals accordingly.

Lets summarize the results of our analysis. Again, a word of caution. Based on the neural circuits of the cerebellum, the inputs to and targets of that structure and what we know of how the entire muscular output control system works, this is what seems to be the most likely functionality of the cerebellum.

The whole cerebellar cortex is continuously involved in controlling your muscular output behavior. That behavior is essentially one output program at a time in a continuous stream of serial events. A large percentage of voluntary movement output actually comes from the cerebellum. If you are now sitting in a chair reading and turning the pages of this book, the action of turning the pages is completely learned and automatic. You don't consciously think "I need to turn the page". Your cerebellum recognizes the need to turn the page in the state of your cerebral cortex and provides the set of motor output patterns that accomplishes the task. Most of your voluntary movement is learned and actually driven by your cerebellum. The cerebellum is continuously providing learned muscular control that allows you to attend to other concerns as you interact physically with your environment.

The operation of the neocerebellum is what allows humans to perform complex voluntary output behaviors built hierarchically. The playing of the piano is a perfect example.

The human neocerebellum is what allows the cerebral cortex to concentrate on displaying the emotion of the music and not the mechanics of the playing. Only the human has this capability and the human neocerebellum makes it possible.

Component #4 - The Hypothalamus and Your Limbic System

Limbic control first appeared around 300 million years ago. This first experiment in hormonal control became the hypothalamus and associated limbic nuclei. These components develop from the basal plate and are output components. The appearance of limbic cortex appeared with reptiles around 250 million years ago. All cortex develops from the alar plate and is an input component.

This state of the limbic system supported dinosaurs until 65 million years ago. The difference between the limbic systems of dinosaurs and humans is the amount of neocortex supported, especially frontal cortex. The frontal cortex is not a part of the limbic system but is heavily involved in its function.

The hypothalamus lives just under the thalamus and is a relatively small neural component composed of many nuclei that perform a wealth of separate and distinct functions. Other main limbic nuclei are the olfactory receiving septal nucleus and the amygdala located at the forward tip of the temporal cortex. Limbic cortex forms the mouth of your cortical balloon and contains the cingulate gyrus and hippocampal formation. Frontal cortex is located forward of the lateral sulcus and makes up one third of all cortex in humans.

What I find most interesting about the limbic system is that, with the exception of the limbic cortical lobe, the limbic system is entirely hardwired. All limbic and hypothalamic nuclei are genetically specified. The neural nuclei that drive your emotions are a reflection of your genes. Your limbic system is determined by one set of genes derived from the combination of available genes from your ancestral pool. Your core personality traits, introverted or extroverted, aggressive or passive, caring or distant, all of the attributes that describe your basic personality are inherited and unaltered since birth.

The hypothalamus regulates your body in response to both your internal and external environments. It reacts to internal state in order to maintain an internal consistency and alters the internal state in times of need. The hypothalamus achieves its control over internal state by driving both the endocrine hormonal system and the autonomic nervous system. Mastery over the endocrine system is accomplished through output to the pituitary and pineal glands located just under the hypothalamus. Control of the autonomic nervous system is accomplished through projections to the brain stem and spinal cord.

The hypothalamus receives input from the midbrain, olfactory septal nucleus, visual tract, amygdala, thalamus, hippocampus, and frontal cortex. It projects to the midbrain, spinal cord, reticular formation, frontal cortex via the thalamus, and pineal and pituitary glands. The output bursts of axonal action potentials to the pineal and pituitary glands produce measured pulses of secretion and not a continuous flow.

Lets take a quick look at the diversity of hardwired limbic functions. The hypothalamus is divided into five regions. The lateral region pattern detects visceral and olfactory input and projects to the midbrain. Stimulation of the lateral area produces the bodily symptoms of anger with the corresponding rise in heart rate and blood pressure. Lesions of this area cause the reverse symptoms and produce an animal that is placid. The preoptic region projects to the pituitary gland to produce gonad-stimulating hormones. The supraoptic region contains four distinct nuclei that control the production of hormones, the circadian rhythm, and parasympathetic heart action. The tuberal region contains three distinct nuclei that control hunger and satiety and produce the emotion of rage when stimulated. The mammillary region controls the sympathetic action of the heart and gastronomic tract. It sends projections to the midbrain and also to the cingulate gyrus through the anterior nucleus of the thalamus.

The septal nucleus receives olfactory input and projects to the hypothalamus. The importance of olfactory input in formulating emotions is much diminished in humans. The amygdala is a well-studied nucleus at the frontal tip of the temporal lobe. Electrical stimulation of various portions of the amygdala can produce total rage or conversely total placidity. The amygdala has been shown to play an important role in memory formation and learning.

The major areas of the limbic lobe are the cingulate gyrus and parahippocampal formation. Both areas are involved in linking memories with emotions such as a picture that makes you feel happy. The hippocampus and parahippocampal gyrus are limbic areas involved in facilitating memory. The strength of your memories is tied to your emotional state at the time of their storage. Many of your most intense memories center on events where you were severely embarrassed or incurred a similarly intense emotion. This is only the extreme example of emotions effect on memory storage. Your emotional state always has a direct effect on the strength of memory storage. This emotional control over memory formation is facilitated by the interconnections between the hippocampus and the rest of the limbic system.

The frontal cortex is intimately involved with the limbic system in shaping your emotional state. Your consciousness of your emotions or feelings are a combination of your input senses detecting what the hypothalamus is doing with your body and the direct input to the frontal cortex from the hypothalamus and limbic nuclei. Humans have the ability to consciously affect their emotional state. "Think happy thoughts" is advice that if followed, directly influences the entire limbic system to change emotional state.

The state of the body is controlled by the limbic system in order to allow a coordinated response relevant to the current environment. When a memory pattern is recognized, limbic cortex causes the limbic system to generate the emotional state associated with that memory. The hippocampus receives currently active memory and projects to the hypothalamus to illicit the corresponding emotional state. This is done for both declarative memories and procedural memories. Your limbic system projects continuously through the thalamus to your entire frontal cortex where procedural memories are stored and directly to your basal ganglia. This frontal cortex input is not concentrated like sensory input but is diffusely provided to all areas of the frontal cortex.

When you access a declarative memory, the limbic system generates an emotional response to the declarative object. That emotional response is input to the frontal cortex where procedural memories are stored and to the basal ganglia. This input affects the selection of procedural memories that drive your behavior and is part of the criteria for the selection of what behavior to choose. If you see a particular food that you do not enjoy, your limbic generated negative response to that food is projected to the frontal cortex and you reject eating it.

The frontal cortex's role in the limbic system is built on top of the older limbic system. As in all neural systems, the older system is not replaced but is augmented and in some ways controlled by the newer cortical system. The frontal cortex is associated with what we term conscious behavior. The implication is that this is logical behavior and superior to emotional behavior. We humans have a short attention span and do not exercise continuous conscious control over our emotions. The old limbic system does not have an attention span. It is always on. If you are not asserting conscious control, your basic personality inherited from your ancestors, drives your behavior.

Component #5 – The Thalamus

The next three components we will cover, the thalamus, cerebral cortex, and basal ganglia, form a tightly interconnected system. These neural components have evolved together to enable your intelligence. Understanding how these three components interconnect and function as a system is critical to understanding how your brain works. We often equate human intelligence to the fact that we have a great deal more cerebral cortex than other animals. The size of your thalamic and basal ganglia neural structures are just as superior to other animals as the amount of cerebral cortex you have inside your skull.

The evolutionary expansion of the cerebral cortical sheet in humans has been matched by an equal expansion of interconnected neurons in the thalamus and basal ganglia. Of these three interconnected neural components, the thalamus is the only one that is completely hardwired. There is no evidence of any learning in the pattern detection capabilities of any thalamic neurons. The fact that the thalamus is hardwired does not lessen its importance. The thalamus plays a central controlling role in supporting the function of the cerebral cortex. The thalamus develops from the alar plate in the growing embryo and is an input type structure.

There are two types of thalamic nuclei, gating type nuclei and reticular type nuclei. All gating type nuclei form identical reciprocal connections with the cerebral cortex called thalamocortical loops. The entire cortex, with the exception of the hippocampus, is interconnected with thalamic gating nuclei via thalamocortical loops.

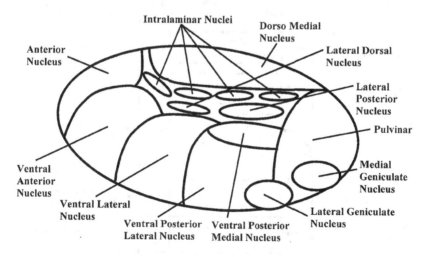

Figure 5-21: Thalamic Nuclei

There are four classes of gating nuclei, sensory input nuclei, limbic input nuclei, motor input nuclei, and cerebral association area nuclei. Reticular type nuclei provide both excitatory and inhibitory control for gating type nuclei. We will briefly discuss the thalamic nuclei that constitute each of these types. (Reference Fig. 5-21 and Fig. 5-22)

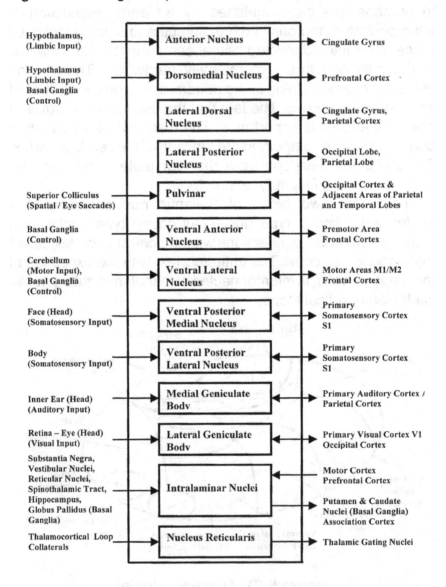

Figure 5-22: Thalamic Nuclei Input and Output

Four thalamic nuclei gate your input sensory channels. Somatosensory input from the head and body is handled by the ventral posterior medial and lateral thalamic nuclei. The medial geniculate nucleus gates incoming auditory input to the primary auditory cortex (A1) in the temporal lobe and the lateral geniculate nucleus gates visual input to the primary visual cortex (V1) in the occipital lobe.

All of these sensory input nuclei share common characteristics. Sensory neural signals are passed to the cerebral cortex for the most part unchanged. This allows the first stage of neural pattern detection in the cerebral cortex to have an undistorted representation of your sensory perception of the external environment. The neural function that does take place in sensory gating type nuclei is inhibition that determines which input signals will be passed to the cortex.

Sensory input nuclei are layered with different types of sensory inputs concentrated in each layer. For example pain and temperature in one layer, smooth skin in another, etc. The various layers of thalamic neurons that represent a single coherent sensory input area, such as a patch of skin, are aligned vertically within the thalamic nuclei.

Two thalamic nuclei gate limbic system input to the cerebral cortex. The anterior nucleus gates limbic input from the hypothalamus mainly to the cingulate gyrus. This connection forms one part of the path that is essential in the formation of memory and learning. This is the largest of the thalamic nuclei and that is an indication of the importance of this function. The dorsomedial nucleus gates limbic input from the hypothalamus to the prefrontal cortex. This limbic input is part of the pattern detection of procedural memories.

Two thalamic nuclei are involved in your voluntary motor control function. The ventral anterior nucleus provides standard thalamocortical loops with the premotor cortex. These thalamocortical loop gating circuits are under basal ganglia control. The ventral lateral nucleus gates motor control input from the cerebellum to the primary motor cortex, M1. These thalamocortical loop gating circuits are also under basal ganglia control.

Three nuclei interconnect via thalamocortical loops with association areas. The lateral dorsal nucleus supports connections with the cingulate gyrus and overlying portions of the parietal cortex. The lateral posterior similarly supports the association cortex portion of the parietal and occipital lobes and the pulvinar thalamic nucleus interconnects with the visual recognition association cortex areas of the occipital, parietal and temporal lobes.

There are two reticular type nuclei in the thalamus, the intralaminar nuclei and the nucleus reticularis. The intralaminar nuclei resemble an extension of the reticular formation seen in the brain stem. As in the brain stem, they lie along the medial long axis of the thalamus and form a Y shaped layer between the gating nuclei. The intralaminar nuclei receive projections from the frontal cortex, basal ganglia, hippocampus and the midbrain. They project to the basal ganglia and widely to the association areas of the cerebral cortex.

The nucleus reticularis, not shown in Figure 5-21, is unlike any of the other thalamic nuclei. The neurons of this nucleus cover the top and sides of the entire thalamus in a thin sheet. The thalamocortical projections from the thalamus to the cerebral cortex and also the reciprocal projections from the cerebral cortex back to the thalamus provide collaterals to this neural sheet of reticularis neurons. This sheet like nucleus receives the entire neural reciprocal traffic between the thalamus and cerebral cortex. This access to all neural connections between the thalamus and cerebral cortex allows the nucleus reticularis to provide a critical control function that is fundamental to the operation of the cerebral cortex.

Nucleus reticularis neurons have extensive dendritic arbors that sample reciprocal thalamocortical communication over a wide area. These widespread dendritic arbors also contain dendro-dendritic synapses with other neighboring nucleus reticularis neurons. The dendritic arbors of nucleus reticularis neurons are synaptically interconnected to form an integrated dendritic lattice that receives all communication from the entire cerebral cortex. The nucleus reticularis forms an interconnected network of neurons that are involved in the widespread inhibition of thalamic gating nuclei. This sheet shaped nucleus appears to integrate the entire cerebral cortex and the entire gating thalamus into one coherent system.

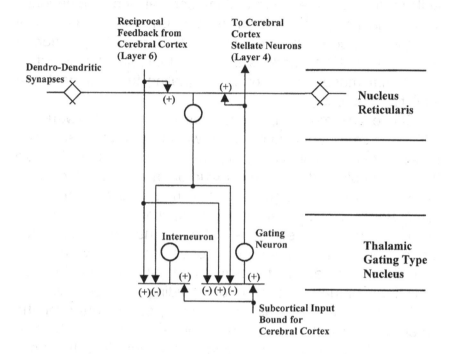

Figure 5-23: Thalamic Gating Circuit (Thalamocortical Loop)

The fundamental neural circuit that makes up the thalamus is a gate that is either open or closed. This thalamic gate is a common design for all of the gating type nuclei. (Ref. Fig. 5-23) The nucleus reticularis neuron has inhibitory control of the gating neuron and its primary function is to close that gate and disable the associated thalamocortical loop.

The axonal projection of the thalamic gating neuron targets stellate neurons in layer four of the cerebral cortex. These stellate neurons project to the spines of pyramidal neurons in all layers of the cortex. The reciprocal feedback from the cerebral cortex comes from pyramidal neurons in the bottom layer six.

These two signals constitute the thalamocortical loop that connects the entire thalamus and cerebral cortex together.

The inhibitory nucleus reticularis is positioned between the thalamus gating neuron and the cerebral cortex and has the ability to inhibit or break this thalamocortical loop. The dendritic interconnections of the nucleus reticularis make it look logically as one interconnected network. The uniquely positioned nucleus reticularis contains the neural circuitry for selecting which thalamocortical loops are signaling the strongest and inhibiting the rest.

One of the most important questions to be answered in understanding how the human brain works is how the association of separate and distinct cortical patterns is supported. How are the word apple and the visual image of an apple related within the human brain? A design mechanism must exist for interlocking disparate areas of cortical pattern recognition into a coordinated whole. The nucleus reticularis is the only neural structure that interacts with all thalamocortical loops. Its extensive dendro-dendritic connected input structure and gate inhibiting function make it the primary candidate for providing this mechanism.

The thalamus acts as an incoming switchboard for the cerebral cortex. In that capacity it performs an inhibiting modulation function on incoming information. The thalamus interacts with the cerebral cortex in a typical winner-take-all neural system to determine what input receives attention. The thalamus plays a much bigger role than just a dumb switch under control of the cerebral cortex. The thalamus has connections to the entire cortex and performs a standard critical function for all areas. We will examine that standard function further but first we need to examine the other end of our interconnected thalamocortical system, the cerebral cortex.

Component #6 - The Cerebral Cortex

Your cerebral cortex is where the patterns that define you and your life are recognized. At birth your cerebral cortex is virtually a blank sheet. You recognize no objects, display no learned behavior, have no memories, and recognize no space. You must build within your cerebral cortex a representation of your world and everything in it. That pattern recognition learning is very intense in the first years of your life and kills a lot of neurons. The learning of new patterns is much less intense in adulthood but is never over. You go on learning and changing with every waking day.

The recognition of previously encountered patterns in the cerebral cortex is referred to as memory. Declarative memories represent objects and space and are contained in non-frontal cortex. Procedural memories represent learned behavior and are contained in frontal cortex. The role of the cerebral cortex is exactly the same in support of both types of memory. We will begin our discussion of the cerebral cortex with its oldest portion, the hippocampus.

The hippocampus is your most ancient cortex and exists in all animals that have any cortex at all. The structure and neural wiring of the hippocampus are consistent across all of the species that contain it. Neocortex evolved after the hippocampus and is dependent upon it for normal operation.

The hippocampus is a rolled up edge of the cortex on the inner side of the temporal lobe. It forms the lower mouth of the balloon like cortical sheet and contains three layers instead of the normal six. The hippocampus more closely resembles the cerebellum than neocortex in both the number of its layers and in its regular neural structure. Hippocampus pyramidal cells, like purkinje neurons in the cerebellum, have long dendritic structures that are crossed by numerous long axons. This matrix like arrangement allows the hippocampus to serve as a major convergence zone for input from the association areas of the neocortex.

The neurons that grow to form the hippocampal formation start growth on the inner side of each hemisphere. The growing mass of neurons expands posteriorly in a vertical circular pattern growing their axonal projections as they migrate. This path of axons becomes the limbic interconnecting fornix. This circular pattern deposits the neural mass as the most medial portion of the temporal lobe where it forms the hippocampus and dentate structures.

The hippocampus continues to develop postnatally and is not operational until some period after birth. The hippocampus is also unfortunately a brain structure that declines linearly with advancing age. Senility or dementia is almost always associated with a loss of memory capability and this is partially due to a loss of neurons in the hippocampus. The hippocampus does not seem to be as plastic as the neocortex. Rats raised in impoverished or enriched environments do not have differences in the dendritic structure of their hippocampus.

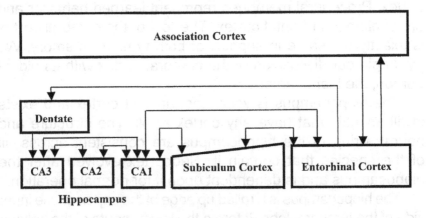

Figure 5-24: Hippocampal System

The dentate nucleus receives input from the associative areas of the neocortex via the entorhinal cortex. (Ref. Fig. 5-24) The dentate nucleus houses granule neurons that send axonal projections widely to the external molecular layer of the hippocampal gyrus. They form mossy tufts with the dendritic arbors of hippocampal pyramidal cells, reminiscent of the neural structures we encountered in the cerebellum. The middle pyramidal layer of the hippocampus consists of two to four layers of pyramidal cells that have their dendritic arbors in the molecular layer and are the only neurons to project axons outside of the hippocampus. These pyramidal neurons support NMDA type dendritic synapses. The internal polymorph layer contains basket cells that receive collateral input from pyramidal cells and target their strongly inhibiting axonal projections to the bodies of pyramidal neurons, again, strongly reminiscent of the cerebellum.

The CA1 area has significant reciprocal connections with the subiculum. The subiculum projects to the entorhinal cortex and thus back to the association areas of the neocortex. The final areas supporting cascaded pattern detection of dentate output, area CA1 and subicular cortex, are the areas that grow the most in percentage as we progress from rats to humans.

Two communication paths are utilized by the hippocampus, the fornix and the alveus. The fornix interconnects the hippocampus with the rest of the brain's limbic system. The alveus provides the reciprocal connection path between the hippocampus and association cortices via the entorhinal cortex. The hippocampus bridges the limbic system and the associative areas of cerebral cortex where memory patterns are recognized. The hippocampus is not reciprocally connected via thalamocortical loops to a gating type nucleus in the thalamus, the only portion of the cerebral cortex not connected in this way.

High-level association cortex projects the currently active pattern to the hippocampus via the entorhinal cortex. The hippocampus projection to the limbic system causes the limbic system to generate the emotional state associated with those memories. The hippocampus is central to the generation of emotions associated with declarative memories and procedural behavior.

The famous patient HM, who had his hippocampus removed, had total recall of everything that occurred prior to his operation. He had no recall of anything that occurred after his hippocampus was removed. The hippocampus is not the storage location for memories and is not required for existing memories to be recognized. That leaves the memory write function as the hippocampal contribution.

The hippocampus pattern detects limbic input in conjunction with the current dominant cortical pattern and provides reciprocal feedback to both the limbic system and active cortical areas. That reciprocal feedback projection indicates "save this pattern with this intensity". That intensity is directly proportional to your emotional state. If you are missing a hippocampus, that intensity is zero. The association cortex projecting the dominant cortical pattern receives no feedback and the dominant cortical pattern is not modified in a manner that allows it to be recognized.

A brain with no cerebral cortex is essentially hard wired. Just to be territorial requires that you have the ability to learn to recognize your territory. To travel any distance from whatever you call home requires you to learn what your home looks like. The primitive frog has cortex but no neocortex. All mammals have limbic cortex, primary sensory areas for somatosensory, sound, visual, and chemical taste, secondary sensory areas for visual and somatosensory, visually related temporal cortex for object recognition, and some prefrontal association cortex. Most of the additional cortex that has been added from early mammals to humans is association cortex. This increase in association cortex is most evident in primates and especially in humans where association cortex is the major form of cortex.

The cerebral cortex is made up of horizontal layers of neurons that communicate vertically. The neuron density in any one layer varies depending on utilization. The vertical structure also varies from somewhat disorganized to highly organized arrays of columnar structures. The neural circuits and overall structure of the cortex are consistent throughout.

The cerebral cortex begins life as a highly over populated sheet of neurons with copious neural wiring. With use, inefficient synapses and neurons are eliminated. The actual final wiring and use of a particular section of cortex is entirely dependent on its input and where it sits in the cascade from sensory input to motor output. Each individual's cortex is unique to his or her development and experience and there is a great deal of individual variability in the functional map of the cortex.

Even though there is a great deal of individuality, we all have the same basic areas with the same basic wiring. The major gyri and sulci are determined by neural interconnections during development and are therefore the same in all humans. The external appearance of the cerebral cortex is similar in all humans.

The cortex grows in the embryo somewhat as it evolved. The cortex that will form the archeocerebral hippocampus grows slowly and partitions into the paleocerebral limbic collar. The edges of the sheet form the mouth as the fast growing neocortex expands upward forming a smooth surface curved balloon at six months. The neocortex first expands anteriorly to form the frontal cortex and then posteriorly to form the parietal cortex. Finally, the cortex expands downward to form the occipital and temporal lobes. This expanding balloon runs out of room forcing the sheet to fold much like stuffing a piece of paper into a small space. This folding greatly increases the surface area of the human cortex as approximately two thirds of the cortex is buried in sulci.

In primates, all neocortical neurons exist before birth. At birth, the dendritic arbors of cortical neurons are quite small. They begin growth after axonal inputs from other cortical areas and subcortical inputs arrive. These early, numerous neurons grow wildly and form extensive dendritic and axonal connections in a somewhat random fashion. One third to one half of these synapses will be lost before adulthood. This weaning of neurons and synapses during childhood ends with adulthood and very few neurons in the neocortex are lost after childhood.

After the growth of cortical axons, the axons are mylinated by specialized glial cells. This myelination takes a number of years and follows a set sequence. Sensory and motor functions are mylinated first, followed by secondary sensory and premotor functions, then declarative association cortex, and finally, the prefrontal association cortex. The prefrontal cortex does not become fully mylinated until puberty.

The pattern recognition tasks that an area of cortex performs are determined by the neural input available to build useful pattern recognition. The area of cortex containing a memory pattern does not expand with use or training, it shrinks. The more familiar the pattern, the smaller the neural area utilized to recognize it and the more strongly those neurons signal when the familiar pattern is encountered.

There is a general flow of pattern detection from the top layers downward to the lower neurons that drive the cortical output to other destinations. Thalamic input is concentrated around layer four with some input to all layers. Stellate neurons receive this thalamic input and project to the dendritic spines of pyramidal neurons. Cortical input is concentrated in layers two and three. Output comes from layers three, five, and six. Layer three targets ipsilateral cortex, contralateral cortex and the hippocampus. Layer five projects to the basal ganglia, pontine nuclei, and in the case of M1, motor neurons. Layer six projects to the thalamic gating type neurons that build thalamocortical loops for that area. Vertically signaling pyramidal neurons project horizontal collaterals to adjacent cortical areas. These horizontal collaterals are concentrated in layers three and five.

Cerebral pyramidal neurons contain spines within their dendritic arbors. These dendritic spines contain NMDA type synapses capable of synchronous gating of input. The greatest concentration of dendritic spines and NMDA synapses is in layers two and three.

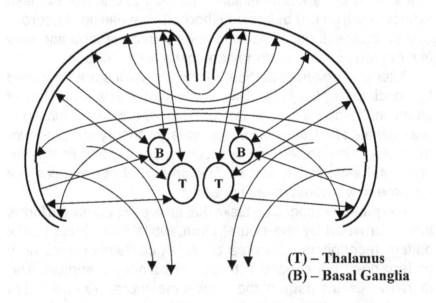

(T) – Thalamus
(B) – Basal Ganglia

Figure 5-25: Cortical Wiring

Large-scale cortical wiring is essentially the same in all humans. (Ref. Fig. 5-25) There is a one to one correspondence between all cortical areas and a gating type neuron in the thalamus. All cortical areas send projections to the basal ganglia and pontine nuclei and have reciprocal connections with the entorhinal cortex supporting connection to the hippocampus and limbic system.

The flow of cascaded pattern recognition in the cortex flows from primary sensory cortex, to secondary sensory cortex, to spatial and object related association cortex, to prefrontal association cortex, to premotor cortex, and finally to actual output in the primary motor cortex. An interlocking chain of cortical areas supports this neural flow. Each link in this flowing cortical chain of pattern detection is reciprocally interconnected with both forward and return multiple neighboring links. The entire cortical chain from secondary sensory areas to behavioral output is tied together with typically three to seven reciprocal connections to neighboring links. In secondary visual cortex, each area is typically interconnected with three to six other ipsilateral areas and from one to three contralateral areas.

This results in an inter hemisphere connection structure that is massive. Short-range ipsilateral connections are contained within individual sulci. Short looping bundles of axons interconnect adjacent gyri ubiquitously. Extremely large bundles of axons dip down into the white matter to interconnect all of the major lobes. The corpus callosum is a massive reciprocal contralateral interconnection path that binds the two hemispheres together. The entire cortex, including the hippocampus, is reciprocally interconnected with its counterpart in the opposing hemisphere.

All of this interconnectivity allows the large sheet of cortex to operate as one integrated neural component. Every area in this chain of pattern detection has the necessary reciprocal feedback to determine if it is part of a winner-takes all cortical pattern detection.

We have covered most of the cortical lobes in some detail due to their role in sensory input or their interaction with subcortical neural components. The only major lobe that we have not covered in detail is the frontal cortex. The frontal cortex has three main areas. Prefrontal cortex is frontal association cortex that houses the patterns of procedural memory. This is the final association area. The most complex pattern recognitions are built here. Representations of possible action, planning, abstract thinking, and behavior are prefrontal functions. This is by far the largest area and makes up the majority of the forward portion of your frontal cortex. This prefrontal cortex and the higher level association areas of the parietal and temporal cortex that reciprocally interconnect with it are the most recent of neocortical areas to expand in human evolution.

Premotor cortex is the equivalent of secondary somatosensory cortex in the parietal lobe. This cortex pattern detects the current active procedural memory and drives the final area, the primary motor cortex, M1. Premotor cortex is involved in the preparation of actual movement. Similar to secondary sensory input cortex, this cortex contains multiple body maps that are combined via pattern detection to drive the actual primary motor cortex.

Premotor cortex contains learned patterns that chain movements together into actions over time. This is coordination of many small movements into a large complex movement. Here are the patterns that throw a ball or swing a racket. The force and velocity of movement are also a part of the control provided by the premotor cortex. At the bottom of the premotor cortex adjacent to the motor output area controlling the mouth and tongue is an area called Broca's area that is involved in speech. Lesions to this area destroy the ability to speak.

Your primary motor cortex, M1, controls individual groups of muscle fibers. Instead of building ever more complex patterns from input, the motor cortex decomposes complex behaviors into ever more granular patterns driving individual muscle fiber groups. Pyramidal neurons in layer five of M1 project directly to motor neurons innervating muscle fibers. M1 receives motor control input from the premotor cortex and from the cerebellum via the ventral lateral nucleus of the thalamus. The input from the cerebellum represents learned motor control patterns.

Behavior is a process that involves the entire cerebral cortex. Patterns of sensory input are recognized and built into ever more complex pattern recognitions. These pattern recognitions are projected to the frontal cortex where they are integrated into pattern detection that includes limbic input. A procedural memory is chosen and the behavior pattern is decomposed into its constituent parts and executed. Procedural behavioral patterns are developed in exactly the same way as declarative perceptual patterns. The frontal cortex is at the end of a long chain of ever more complex pattern recognitions. The pattern recognitions that achieve dominance in the frontal cortex determine behavior. The behavior of humans is complex and you have an amazing amount of frontal cortex to house that complexity.

The cerebral cortex is where you build the patterns of your experience. It does not operate alone and requires the proper functioning of the thalamus in order to serve as the repository of memory. A memory is represented in the cerebral cortex as a complex pattern of firing pyramidal neurons that spans large and discontinuous areas of cortex. A memory is recognized when its unique complex pattern achieves dominance within the cortex. Combining what we know about the cerebral cortex and supporting thalamus we will now analyze how memory appears to be implemented in the cerebral cortex.

Each independent vertical area of cerebral cortex performs pattern detection as a whole. In other words, all of the pyramidal neurons in a vertical area signal essentially the same pattern recognition. Different layers of pyramidal neurons project to different neural targets. If the cerebral cortex did not perform pattern detection as a vertical whole, the various neural systems that receive cerebral input would get different patterns of inputs. In order for the brain to operate as an integrated whole, the state of pattern recognition in the cerebral cortex needs to be projected equally to all targets. This has the effect of making the memory pattern in the cortical sheet two-dimensional.

Memory recognition is literally constructed within the cortex from continuous waves of input patterns that wash across the cortex from sensory input through declarative memory and on to procedural memory. A memory recognition pattern is built into the cortex through cascaded pattern recognition across the entire cortex. Each level of cascaded cortex memory recognition requires input that is synchronous in order to be recognized. Memory pattern recognition is implemented by synchronized waves of pattern input that wash across the cortex.

For a complex memory pattern to be recognized, it must win in a winner-take-all contest with all of the other possible patterns currently playing within the cortex. This contest first starts locally and ends up as a winner-take all memory pattern involving large discontinuous areas of cortex.

Local contests start out with dominance of particular vertical groups of pyramidal neurons over neighboring vertical groups of neurons. The numerous horizontal collaterals of pyramidal neurons predominately from layers three and five serve as input for these competitions. These local contests are refereed by stellate basket neurons that detect when neighboring areas of pyramidal neurons are firing with greater strength and shut down their area by targeting the cell bodies of their pyramidal neurons.

Each area of cortex projects the strength of its pattern recognition to the thalamus. All of these signals pass through the nucleus reticularis that provides the next level of winner-take all implementation. The nucleus reticularis consists of a one-neuron thick layer that covers the entire thalamus. All thalamocortical loops pass through and signal to the nucleus reticularis. Each nucleus reticularis neuron is strongly inhibiting and has the ability to shut down thalamic gating neurons in a very localized area. The dendritic arbors of reticularis neurons are extensive and intersect a large number of thalamocortical loops. The dendritic arbors of nucleus reticularis neurons also contain dendro-dendritic electrical gap synapses that bind the sheet of nucleus reticularis neurons into a coordinated whole.

The nucleus reticularis sheet of neurons has the ability to inhibit the firing of every thalamic gating neuron. This gives the nucleus reticularis inhibitory control over literally every thalamocortical loop that binds the thalamus and cortex together. The function provided by thalamocortical loops is what facilitates cascaded pattern detection and thus memory recognition within the cortex. How this works in detail we will examine shortly.

This sheet of inhibiting neurons disables all but the most strongly signaling thalamocortical loops and causes those strongly signaling thalamocortical loops to be synchronous. The neurons in the nucleus reticularis form a matrix that serves to interlock individual cortical areas that signal strong pattern recognition into a larger collection of strongly signaling areas to form a larger synchronous pattern. This larger pattern suppresses lesser-signaling patterns and serves as synchronous input to the next level of more complicated pattern recognition. The result is a synchronized winner-take-all implementation across the entire cortex.

Pattern detection cascades forward in ever increasing complexity. At each level within this cascade, actually at each individual pyramidal neuron, inputs that constitute patterns to be recognized must be synchronous in order to be recognized. I am going to postulate that the neural circuitry of the thalamocortical loop gates pattern detection within the cerebral cortex and causes the cortical areas that are part of a forming memory to synchronize. We will perform circuit analysis of the thalamocortical loop to understand how it appears to work. (Ref. Fig. 5-26)

Understanding how thalamocortical loops facilitate the operation of memory is critical to an understanding of how your brain works. Please refer to the neural circuit depicted in figure 5-26 in the following circuit analysis. The major neurons that make up the thalamocortical loop are the pyramidal and stellate neurons of the cerebral cortex and the gating type and nucleus reticularis neurons of the thalamus.

The dendritic trees of cerebral pyramidal neurons contain NMDA type synapses on all of their dendritic spines. A magnesium plug that renders the synapse inactive normally blocks NMDA synapses. An enabling excitatory input must be received on the dendritic spine in order to remove this magnesium plug and enable the pyramidal NMDA synapses to be active. This NMDA gating type synapse allows input to cerebral pyramidal neurons to be controlled. This is the neural mechanism that facilitates cascaded pattern detection.

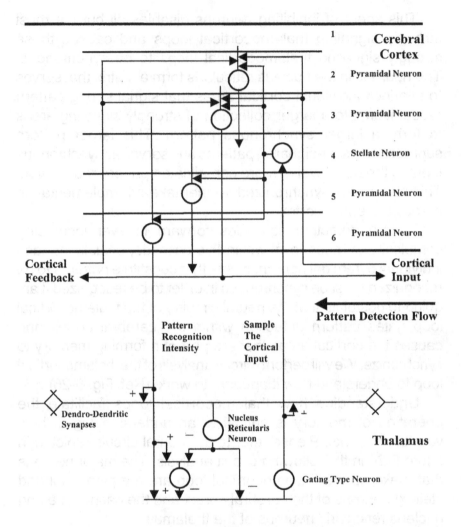

Figure 5-26: Synchronization Circuit Thalamocortical Loop

Stellate neurons project excitatory input to the dendritic spines of all cerebral pyramidal neurons. This excitatory input serves as the required enabling input that removes the magnesium plugs and allows the pyramidal neurons to pattern detect their input. The thalamic gating neurons project excitatory input to all cerebral stellate neurons. This excitatory input causes stellate neurons to fire and enable pattern detection by cerebral pyramidal neurons.

The thalamic gating type neuron projects to all stellate neurons and a firing thalamic gating type neuron always causes the stellate neurons to fire. Stellate neurons project to the NMDA dendritic spines of pyramidal neurons in all levels. That excitatory input signal from the stellate neurons removes the magnesium plugs associated with NMDA synapses and allows the pyramidal neurons to sample their input. If the thalamic gating type neuron is not active, the stellate neurons remain off and the magnesium plugs continue to block the synapses from working. If the thalamic gating type neuron is not active, this cortical area is effectively turned off and no pattern recognition is allowed.

The nucleus reticularis neurons project strongly inhibiting input to the thalamic gating neurons. The thalamic gating type neuron is under the inhibitory control of the nucleus reticularis neuron. The nucleus reticularis neuron has the ability to disable the thalamic gating type neuron and therefore has control over when the cortical pyramidal neurons are allowed to pattern detect incoming input. This inhibitory control also gives the reticularis neurons the ability to shut down the gating neurons at any time and disable pattern detection by cerebral pyramidal neurons. The large, dendro-dendritic connected dendritic arbors of nucleus reticularis neurons allows the nucleus reticularis neuron to sample thalamocortical loop activity over a very large area in order to control the pattern detection of each area of cerebral cortex.

Synchronization of thalamic gating neuron firings by the thalamic nucleus reticularis will have the effect of causing all of their collective targeted pyramidal neurons in the cerebral cortex to sample their inputs synchronously. Timing this sample input signal with the arrival of synchronous input from upstream pattern detection cortex (Cortical Input) makes for a very efficient cascaded pattern detection system. Remember our flash bulbs in a dark football stadium. The nucleus reticularis also has the ability to synchronize discontinuous areas of signaling cortex by synchronizing the firing of the thalamic gating neurons that support the discontinuous areas.

We will now examine a complete pass of pattern recognition for an individual vertical area of cerebral cortex. An incoming wave of synchronous input approaches. This input wave is matched by a wave of active thalamocortical loops supporting the vertical areas producing this synchronous input. Nucleus reticularis neurons detect this incoming input wave of active thalamocortical loops prior to its arrival at the vertical area of cerebral cortex and this causes the nucleus reticularis neurons to stop inhibiting the thalamic gating neurons. The thalamic gating neurons fire and their output propagates to cortical stellate neurons in layer four and causes them to fire. The axonal outputs of the stellate neurons target the dendritic spines of all of the pyramidal neurons in the vertical cortical area. This input removes the magnesium plugs of the NMDA synapses and allows them to receive the incoming wave of synchronous input that arrives precisely on time.

All pyramidal neurons fire with an intensity that matches their level of pattern recognition for the incoming wave of synchronous input. Stellate basket neurons pattern detect which local areas of cerebral cortex are firing with the greatest intensity and shut down weaker signaling areas. The pyramidal neurons in layer six signal their intensity back to the thalamic gating neurons. These signals pass through and signal to the nucleus reticularis neurons' dendritic arbors. If this intensity is not as strong as other vertical areas of cerebral cortex, the nucleus reticularis neurons inhibit the thalamic gating neurons. The loss of the thalamic gating input causes the stellate neurons to stop firing. This allows the magnesium plugs to block the synapses on the dendritic spines of the pyramidal neurons and thereby shut down this vertical area of cerebral cortex. If the intensity is strong enough, the thalamocortical loops remain enabled.

Pyramidal neurons in level three of the vertical area signal their level of intensity to following areas of cerebral cortex, to the same area of cortex in the opposite cortical hemisphere, to the hippocampus via the entorhinal cortex, and also signal that same intensity as reciprocal feedback to the areas of cerebral cortex that provided the synchronous wave of input.

The collection of pyramidal neurons in vertical areas that remain firing produce the synchronous input wave for the following areas of cerebral cortex and the process continues. After the next cycle of pattern detection of this output wave, the following areas of cerebral cortex that are a part of the forming memory provide cortical feedback to the vertical area. Feedback is also provided by the opposite cortical hemisphere and by the hippocampus via the entorhinal cortex. Positive intensity feedback (Cortical Feedback) to the pyramidal neurons of our vertical area of cerebral cortex indicates that the cortical pattern being recognized remains a part of a forming memory pattern. Positive feedback from the opposite cortical hemisphere indicates agreement in pattern recognition between the left and right cortical hemispheres. Feedback from the hippocampus reflects the current level of emotional state. Lack of feedback from the hippocampus will prevent permanent modification of the current memory pattern.

All of this positive feedback will cause an increase in the firing intensity in the vertical area of cerebral cortex. That firing intensity increase, due to positive feedback, will be detected by the dendritic arbors of nucleus reticularis neurons and a final round of inhibition will occur. This will have the effect of shutting down all of the thalamocortical loops that do not receive positive feedback and are therefore no longer part of a forming memory pattern.

The cycle is complete. The thalamocortical loops of vertical areas of cerebral cortex that are enabled remain enabled. Another wave of synchronous input arrives and starts the process again by causing the nucleus reticularis neurons to allow all of the thalamic gating neurons to fire again. The vertical areas that remain active are a part of a successful memory pattern and have an advantage in the next cycle of synchronous wave input pattern detection due to the positive feedback they are receiving.

In order to be more human friendly, we will adopt a visual projection system analogy to describe how the declarative memory system works. There are three major components involved in the formation and access of declarative memory, the cerebral cortex, the thalamus, and the hippocampus. The cerebral cortex is a storage projection screen, the thalamus is the projector, and the hippocampus determines the intensity with which an image is stored in the storage projection screen. These are not your normal movie projectors and passive projection screens. The cerebral cortex storage screen receives waves of input that roll across all of its areas. Those areas signal to the projector the strength of their pattern recognition for that incoming input.

The thalamus projector has an independent lens for each area of cortex, the thalamocortical loop. Over time, this storage projection screen has seen and stored an amazing number of images. The most strongly stored of those images are very clear and tight and signal strongly when activated. The weakest of them are very fuzzy and diffuse and signal less strongly when activated. Each area of the screen signals to the projector thalamus the clarity of its current image.

The task of the projector thalamus is to facilitate the building of a larger synchronous image from its constituent parts. The building of this larger image in the storage projection screen propagates as a wave from sensory input to sensory association areas, to prefrontal association areas, to output. The thalamus projector facilitates this wave by providing control through its nucleus reticularis sheet to every thalamocortical loop lens.

The function provided by each loop lens is as follows. When an input wave arrives, stop inhibiting the thalamic gating neuron to allow the cerebral cortex area to pattern detect the synchronous input. This sample input enabling signal is synchronous to all thalamocortical loops involved. The cerebral cortex signals the intensity of its pattern recognition of this input. If that intensity is less than the intensity of other thalamocortical loops the lens can detect, the nucleus reticularis inhibits the gating type neuron. If that intensity is equal to or greater than the intensity of other thalamocortical loops, the gating type neuron will remain uninhibited and the cerebral cortex projections to the next levels of cerebral cortex remain on. Feedback from areas of the projection screen that are a part of the winning image help to sharpen the overall image. The final winning image propagates through the cortical storage screen facilitated by the thalamic projector that serves to focus and synchronize the final image.

The final component of this system, the hippocampus, receives most of its cortical input from the higher areas of association cortex. These are areas of the cerebral storage screen that indicate that they are part of a winning overall pattern. The hippocampus is linked to the limbic system and asks the question "Do you care?" The hippocampus signals back to the cortex storage screen the strength of the emotional response to the current dominant memory pattern. That pattern will be stored with that strength. If you have no hippocampus, you have zero interest in storing new patterns.

Lets examine how you visually recognize an apple. You start with a saccade to an apple on the table before you. Your entire visual system has evolved to provide the primary visual cortex (V1) with exactly the kind of input that is optimal for cortical pattern recognition, a series of static synchronous images that can start the wave of visual pattern recognition. The neurons of the retina are inter-linked by electrical gap junctions that cause them to fire synchronously. This synchronous visual input to the thalamus causes the thalamic gating neurons that receive significant input to fire and pass that input on to V1. Raw synchronized visual input reaches the primary visual cortex as a collection of concentric circle detections and primary color differences for all of the retinal fields.

Primary visual cortex columns check each and every retinal field for every possible line orientation. Those columns that detect the pattern for which they are wired, signal strongly to the thalamus, the contralateral equivalent cortical column, and ipsilaterally to the next areas of visual pattern detection. Stellate basket cells within the primary visual cortex limit the number of columns signaling to the ones with the strongest match. The nucleus reticularis of the thalamus receives these signals via thalamocortical loops and shuts down all but the strongest signaling loops. All loops that are signaling strongly oscillate together with each incoming static image.

At this stage, all of the primary visual columns that detect the small line segments comprising the apple are signaling strongly and synchronously. Therefore, their forward propagation projections to secondary visual cortices are synchronous and this allows them to be recognized by the next level of visual pattern recognition. The nucleus reticularis stops inhibiting the gating type neurons of the next level of visual pattern detection in unison and in time to pattern detect this synchronous input. The next level sensitive to larger line segments receives synchronous input containing the tiny line segments and those that recognize learned patterns signal strongly and the process is repeated. In this way the primary input is successively built into larger and more complex representations. This same process occurs for V2, V3 to V5, and on to the posterior temporal lobe where learned patterns for objects are stored.

At this point the inputs to the object recognition posterior temporal lobe are pretty strongly indicative of the pattern of an apple. You will continue to stare at the apple to allow a series of synchronous static visual inputs to wash through your visual cortex until the apple object is recognized. Further saccades to different areas of the apple will also enhance its pattern recognition.

The pattern for an apple is contained in a very large number of neurons. The pattern for an apple started out large and diffuse the first time you encountered one. Each encounter with an apple over your entire life has further modified your cortical pattern for the object apple. It does not matter how big it is or at what angle you see it or what color it is. You have learned the pattern distribution that indicates apple and with this learning the pattern for apple has actually shrank from a large diffuse pattern to a tight well-defined pattern whose individual neurons signal strongly when an apple pattern is encountered. As the center of the apple pattern began to signal more intensely with learning, the edges of the apple pattern that signaled less strongly were dropped by the nucleus reticularis. So the pyramidal neurons in the cortical area that recognizes the object apple fire strongly and we proceed.

The nucleus reticularis performs its task again, measuring the cortical response to the outward propagating wave from the apple object to shut down non-comparing areas. When this wave reaches a higher level of declarative cortex it causes pattern recognition in a cortical area that contains the pattern for the word "apple". In this way the cortical patterns that represent the visual object, the word apple both spoken and written, the taste of an apple, and the smell of an apple are all caused to light up and signal synchronously in your storage projection screen. This process is amazingly fast. You know an apple when you see it.

Cortical pyramidal neurons adapt to the input they receive. They modify their dendritic wiring in response to patterns that cause them to fire and signal recognition of patterns they have seen before. That is all they do. It is enough. It is enough to store a record of your entire life. The cerebral cortex, your storage projection screen, stores a tremendous number of patterns as you live your life. Each stored with a different strength and slightly modified each time it is recognized.

The cerebral cortex is one brain component. It is a massively parallel device. Only one large pattern at a time achieves dominance within the cortex and thus directs behavior. This overall pattern is quite complex. It is not just the collection of synchronously signaling cortical areas that signify an apple. It contains a neural representation of all of the other objects within your environmental space and also a representation of that space. This example of pattern detecting the image of an apple is representative of how declarative memory works. In order to understand how procedural memory works, we are going to have to include the function of another brain component, your basal ganglia.

Component #7 – The Basal Ganglia

The basal ganglia system is the last of our three interconnected brain components that enable human intelligence. The thalamus projector and cerebral cortex storage projection screen implement your declarative memory system. The basal ganglia are an integral part of your procedural memory system that drives all of your behavior.

The control of neural output is fundamentally different from the receipt of neural input. The control of neural output is serial in nature. You only do one thing at a time. From all of your behavioral choices, you choose one. If that choice involves skeletal muscle activation, the procedural memory pattern that signifies that choice activates pattern recognition in premotor and motor cortex to accomplish that activity. Allowing multiple procedural memory output patterns to be active at the same time is obviously not going to work. One output pattern in each instant of time is required if pattern detection by output neurons is going to unambiguously drive behavior.

If you take a second to glance around, you will recognize a variety of objects. These objects all cause different declarative patterns to light up within your cerebral storage projection screen. Each of the cortical area patterns associated with the objects achieves dominance as your eyes saccade to them in scanning your environment. Your attention and focus wanders among the competing object neural patterns. The separate collections of thalamocortical loops associated with these objects remain on and viable as long as the object is near, even if the object never achieves central focus. The association cortices of non-frontal cortex are a dancing pattern of available input to the frontal cortex. The storage projection screen that makes up the frontal cortex is not a cascade of flickering synchronous patterns from competing inputs but a strong single pattern that evolves with time.

I am going to assert that the maintenance of one and only one output pattern at a time and the flow of that pattern in real time to effect smooth, coordinated behavior is the function that the basal ganglia perform. Now we will explore why I assert this function for the basal ganglia.

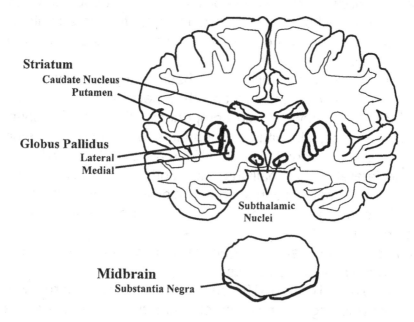

Figure 5-27: Basal Ganglia

The basal ganglia system is made up of nuclei that span from the brainstem substantia negra to the striatum (Ref. Fig. 5-27). The input striatum consists of two physically distinct nuclei, the caudate nucleus and the putamen. These structures are neurologically identical with vast numbers of individual medium spiny neurons serving as both input and output in a virtual single neuron sheet like structure. The caudate nucleus consists of a head under the frontal cortex, a body that proceeds posteriorly under the parietal cortex and a tail that curves down to lie close to temporal cortex. The caudate receives input from the entire neocortex. The putamen makes up the outer portion of a lens shaped structure that includes the globus pallidus nuclei. The putamen is concerned with skeletal muscle control. The striatum acts as a single neuron layer of medium spiny neurons that receives input from motor related and association type cerebral cortices.

The globus pallidus forms the inner portion of the putamen capped lens and the subthalamus lies just under the thalamus. Both the striatum and globus pallidus nuclei support strongly inhibitory projections. The midbrain substantia negra contains dark masses of neurons that utilize the neurotransmitter dopamine in their projections.

The nuclei that comprise the basal ganglia evolved and grew separately as the function they perform changed with the addition of newer brain components. Parts of the basal ganglia are very old with the midbrain substantia negra being the oldest. The medial globus pallidus appeared next followed by the lateral globus pallidus and subthalamic nucleus. The striatum is intimately connected to and developed in conjunction with the cerebral cortex. The putamen controls skeletal output and appeared with early motor and somatosensory cortex. The caudate is the latest basal ganglia addition and services the neocortical association areas.

These nuclei grow at different times from different structures in the developing embryo. The striatum grows from the same embryonic neural area that produces the cerebral hemispheres. As the cortical hemisphere expands it draws the caudate tail backward as the parietal cortex forms and then down and forward as the occipital and temporal lobes are formed. This leaves the caudate tail in its characteristic downward "C" shape lying over the hippocampus. The globus pallidus develops from the same embryonic neural area as the thalamus.

These nuclei are interconnected as a system to effect output control. (Ref. Fig. 5-28) We will describe each of these nuclei and their roles starting with basal ganglia output and working back to basal ganglia input. The medial globus pallidus projects to the thalamic gating type neurons that form thalamocortical loops with the entire frontal cortex. These medial globus pallidus projections are strongly inhibitory and their function is to disable the gating type thalamic neuron and shut down the associated thalamocortical loop. A firing medial globus pallidus neuron effectively shuts down the frontal vertical cerebral area the gating neuron and associated thalamocortical loop supports.

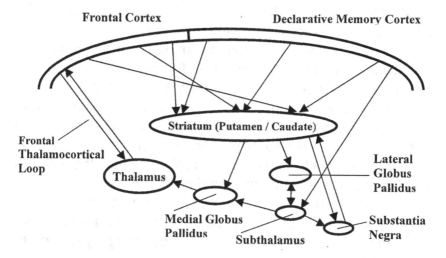

Figure 5-28: Basal Ganglia Interconnections

The inhibitory axonal synapses of medial globus pallidus neurons terminate on the three gating type thalamic nuclei that support thalamocortical loops with the frontal cortex. The ventral lateral thalamic nucleus supports the motor cortex, the ventral anterior thalamic nucleus supports the premotor cortex, and the dorsomedial thalamic nucleus interconnects with the prefrontal cortex. Medial globus pallidus projections to these three thalamic gating nuclei give it inhibitory control of all of the thalamocortical loops that support the entire frontal cortex.

The lateral globus pallidus and subthalamus form a neural circuit that provides medial globus pallidus control. These two nuclei combine to turn on medial globus pallidus neurons when a wave of declarative input arrives at the frontal cortex and then turn them off when the wave passes. This has the effect of providing a disabling window to frontal thalamocortical loops when they would have been enabled by the nucleus reticularis.

The striatum's function is to override all of this inhibitory control. It projects strongly inhibiting input directly to the neurons of the medial globus pallidus and shuts them down. Firing medium spiny neurons have the effect of removing the disabling window to frontal thalamocortical loops, which allows them to operate exactly like declarative cortex.

The medium spiny neurons that make up the striatum are normally in an off state and silent. In this off state the medium spiny neuron has a shunting current that locks it down and makes it very hard for input to excite the neuron to a firing state. It takes a very large number of synchronous excitatory inputs on the medium spiny neuron's approximately 10,000 dendritic inputs to overcome this shunting current and cause the medium spiny neuron to fire. Once the medium spiny neuron does fire, the shunting current disappears and the medium spiny neuron enters an up state. In this firing upstate, a much smaller number of synchronous inputs will cause the medium spiny neuron to continue firing.

The putamen is the striatal structure utilized for motor behavior and it receives a large portion of its input from premotor and secondary somatosensory cortex. Areas of the putamen are dedicated to controlling neural output to specific groups of muscle fibers. Each area of the putamen receives input from primary motor, premotor, primary somatosensory and secondary somatosensory cortex for the muscle fibers it controls. Each area of the putamen receives input from sensory input and motor output cortical areas that are reciprocally interconnected.

The caudate nucleus is physically shaped to receive the massive cortical input from the association areas of the neocortex. It receives input from both declarative and procedural association cortex as a precise two-dimensional linear map over its entire input surface. Each caudate nucleus location receives input from reciprocally connected areas of declarative and procedural cortex. The caudate provides control for all output behavior and is heavily interconnected with the highest levels of association cortex.

There are three primary inputs to the striatum. First, there is the massive output from cortical layer five. These overlapping projections of the declarative and procedural storage projection screens are precisely aligned in a two-dimensional input array to the striatum. Secondly, there is limbic input to the striatum that comes via the intralaminar nucleus of the thalamus. This input combines input from the hippocampus, amygdala and reticular system and impacts the delivery of dopamine. The third major type of input comes from the substantia negra and delivers dopamine to the dendritic spines of striatal medium spiny neurons.

The substantia negra is actually two nuclei stuck together. The first nucleus is an old portion of the basal ganglia that participates in the control of saccadic eye movements. The second substantia negra nucleus contains large numbers of neurons that employ dopamine as their axonal neurotransmitter. This structure reciprocally interconnects with the striatum and delivers dopamine involved in the learning of output control to the striatum and frontal cortex. Remember our berry bushes.

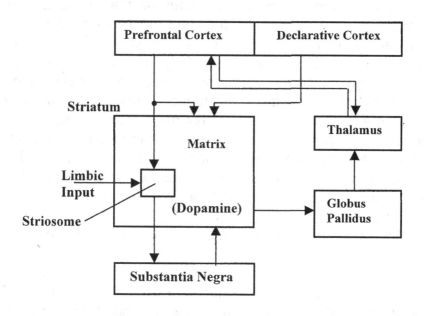

Figure 5-29: Dopamine System

Striatal dopamine delivery occurs via substantia negra axonal synapses to the spines of striatal medium spiny neurons. The neural system that drives this dopamine input is contained within the striatum. There are patches of medium spiny neurons within the striatum called striosomes that drive this dopamineric system. Approximately 5% of medium spiny striatal neurons are contained within striosomes. (Ref. Fig. 5-29) The striatum is a layer of neurons scrunched into a three dimensional physical mass made up of an array of matrix compartments. Each matrix compartment contains a striosome that controls the delivery of dopamine to the medium spiny neurons that make up the matrix.

Striosomes form a regular lattice throughout the striatum. The striatum is a logical sheet of medium spiny neurons that is physically a three dimensional lattice of matrix compartments each containing a striosome. The dendritic arbors and axonal collateral connections of medium spiny neurons within striosomes and matrix compartments are restricted to the compartment that contains them.

Inputs to the medium spiny neurons that make up the striosomes originate from the prefrontal cortex and the intralaminar nucleus of the thalamus. The output of the medium spiny neurons within the striosome targets the substantia negra directly. The striosome controls striatal learning by determining the dopamine input to the matrix that contains it. Limbic input to striosome medium spiny neurons is a major factor in determining dopamine output. The amount of dopamine neurotransmitter released to the matrix areas is a function of positive or negative reward. Increases and decreases in dopamine are associated with causing changes in the striatum, what we term procedural learning.

Before we get into the details of how the basal ganglia control behavior, a quick look at a high level description of the basal ganglia's function. The basal ganglia are an integral part of and have inhibitory control over each and every frontal thalamocortical loop. The basal ganglia service the entire frontal cortex and have the ability to select which thalamocortical loops will be allowed to be operational.

Firing striatal medium spiny neurons remove the disabling of gating neurons and the frontal thalamocortical loops they control. This allows the frontal thalamocortical loop to function under the control of the nucleus reticularis. The pattern of input that causes striatal medium spiny neurons to fire is the current dominant declarative memory pattern lighting up your declarative storage projection screen. This winning declarative pattern causes the striatum to not disable the gating neurons and thalamocortical loops of frontal cortex areas that will pattern detect this input.

Areas of the striatum that do not receive excitatory input from a winning declarative pattern will remain off. The lateral globus pallidus and subthalamus combine to provide a disabling window to the gating neurons and thalamocortical loops controlled by these striatal neurons. This prevents pattern detection in areas not receiving significant declarative input. Only frontal thalamocortical loops that are allowed to operate by the basal ganglia can be a part of a forming procedural memory. Only procedural memories that are allowed by the basal ganglia can drive behavior.

Have a close look at the frontal thalamocortical loop in Figure 5-30. The frontal thalamocortical loop looks exactly like declarative memory thalamocortical loops with one exception. The frontal thalamic gating neuron has an input from the medial globus pallidus that is strongly inhibitory. Frontal thalamocortical loops have two strongly inhibitory control neurons, the medial globus pallidus neuron and the nucleus reticularis neuron. The firing of the medial globus pallidus neuron will inhibit the thalamic gating neuron from firing and disable the thalamocortical loop regardless of what the nucleus reticularis neuron does.

If the striatum medium spiny neuron fires in response to an incoming wave of declarative input, the medium spiny neuron's inhibitory projection to the medial globus pallidus neuron prevents that neuron from firing and control of the frontal thalamocortical loop by the nucleus reticularis is allowed. The firing of the striatal medium spiny neuron has the effect of stopping the inhibition of the frontal thalamocortical loop. Only frontal cortical areas that are controlled by firing medium spiny neurons are allowed to pattern detect declarative memory input.

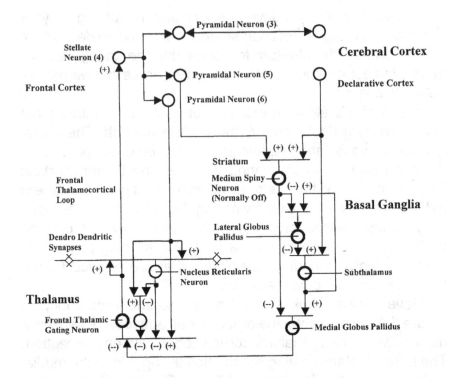

Figure 5-30: Basal Ganglia Loop

The lateral globus pallidus and subthalamus neurons form a control circuit that impacts medial globus pallidus neurons not inhibited by firing striatal medium spiny neurons. The subthalamus receives input directly from the cerebral cortex and can detect an incoming wave of declarative memory. The subthalamus neuron always fires in response to an incoming excitatory wave of declarative input and the neuron's excitatory output is what causes the medial globus pallidus neuron to fire.

Once the incoming wave of declarative memory has passed, the subthalamus excitatory input to the lateral globus pallidus neuron causes it to fire and inhibit the subthalamus, shutting down the medial globus pallidus. The lateral globus pallidus and subthalamus control circuit is now ready for the next wave of declarative memory input. The net effect of causing the medial globus pallidus neuron to fire when a wave of declarative input arrives and stop firing when it has passed, is a disabling window that disables the thalamic gating neuron for this period. All thalamic gating neurons not controlled by firing medium spiny neurons are prevented from enabling their supported frontal cortical area.

In frontal thalamocortical loops controlled by firing medium spiny neurons, the nucleus reticularis detects an incoming wave of active thalamocortical loops. The nucleus reticularis stops inhibiting the thalamic gating neuron, which allows the gating neuron to fire and cause the frontal cortex stellate neurons to fire. This removes the magnesium plugs of the NMDA synapses of the frontal cortical pyramidal neurons and allows them to sample the declarative memory input that arrives precisely on time.

Frontal pyramidal neurons signal their level of pattern recognition for this declarative input back to the nucleus reticularis, the thalamic gating neuron, and to the basal ganglia medium spiny neuron. The nucleus reticularis suppresses all but the strongest signaling thalamocortical loops and sharpens the image as it does for declarative memories. The striatal medium spiny neuron is now receiving the currently active declarative input and the frontal cortex response pattern to that input.

Basal ganglia medium spiny neurons pattern detect declarative input and procedural response and select one procedural memory based on their pattern recognition learning facilitated by dopamine. Only the medium spiny neurons associated with the winning procedural memory will remain on. All but one procedural memory pattern have their frontal thalamocortical loops disabled by the basal ganglia neural circuit.

The active signaling procedural memory pattern is of sufficient strength to cause the firing up state striatal medium spiny neurons to continue firing. This on state will be held by the frontal cortex procedural memory pattern even when the declarative cerebral input that caused it is removed. Frontal cortex procedural memories that are turned on can be held on until the behavior they drive is completed. Only frontal cortex patterns that are enabled by the basal ganglia are allowed. All others remain inhibited.

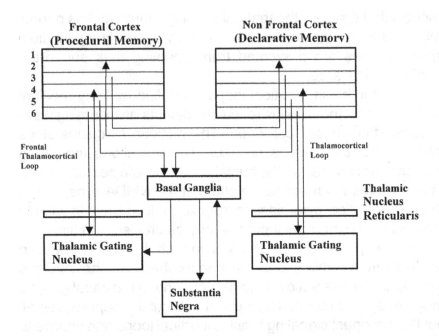

Figure 5-31: Basal Ganglia System

The basal ganglia system has complete control over the thalamocortical loops of the frontal cortex. (Ref. Fig. 5-31) It takes a large number of synchronous inputs from declarative cortex to cause the neurons of the striatum to jump to their up state. Only the synchronous patterns of recognition that light up in your declarative cerebral storage projection screen have the possibility of exciting their particular area of striatum. The striatum is performing pattern recognition on the inputs of the currently dominant declarative memory, the objects and space you have recognized in your environment.

The basal ganglia function is the selection of an output response. One output response from perhaps a myriad of possible procedural memories that could possibly be selected. That process of procedural selection is completely a function of pattern detection within the basal ganglia. Procedural memories are modified by usage just as declarative memories are modified. Behavior however is serial in nature and we do not acquire new behavior nearly as quickly as we acquire declarative memories. Behavior can change quickly and completely based on feedback. If you eat something that tastes really bad, you do not eat it again. The procedural memory associated with eating is not significantly modified; the basal ganglia have been modified to choose another behavior when that food is encountered. That modification of the basal ganglia requires the action of dopamine.

The basal ganglia select one procedural memory that is then utilized to drive behavior. The basal ganglia forms parallel, identical controlling loops with the entire frontal cortex and performs exactly the same function for the primary motor cortex, the premotor cortex and prefrontal cortex. That function ensures that one and only one output procedure is implemented at one time in all frontal cortex areas.

Procedural memory recognition works exactly like declarative memory recognition. The nucleus reticularis and thalamic gating neuron provide an enabling window that gates pattern detection in the cerebral cortex. The pyramidal neurons in level six project their level of pattern recognition back to the gating neuron and provide collateral input to the nucleus reticularis. The nucleus reticularis inhibits all but the strongest signaling areas of frontal cortex and causes those areas to be synchronous.

A fundamental capability of the basal ganglia is the maintenance of an up state that allows procedural memories and their implementation to remain active until the specified behavior is accomplished. Behavior is based on declarative and limbic input but it must be divorced in time from that input. Procedural memory input to the basal ganglia will hold a set of thalamocortical loops in the on state as long as required to accomplish the goal of the behavior. The procedural memories being enabled are not a series of discontinuous snapshots. They are more like a movie. One flows into another in a smooth transition that drives your behavior in a smooth continuous flow.

In our discussion and speculation about basal ganglia function, we have not postulated any role for the basal ganglia in the formation or storage of procedural memory patterns. Procedural memories stored in associative prefrontal cortex are stored in exactly the same way that declarative memories are stored, enabled by hippocampal limbic input. This analysis predicts that our famous patient HM has not stored any procedural memories since his operation. His behavioral repertoire is fixed at his pre operation capability.

Declarative input causes striatal medium spiny neurons to fire and a number of possible prefrontal procedural memories to light up. The basal ganglia choose one for implementation. This selection process can be altered through the action of dopamine to choose different behavior. Substantia negra production of dopamine is controlled by striatal striosomes that receive procedural memory and limbic input. Control of dopamine production allows the basal ganglia to make behavioral choices. The ability to hold frontal thalamocortical loops enabled allows the basal ganglia to control the timing of behavior. The basal ganglia are controlling your behavior, not your frontal cortex.

In our visual projection system analogy for declarative memories we designated the cerebral cortex as your storage projection screen and the thalamus as your multi-lens controlling projector. Continuing with this analogy, the function of the basal ganglia is to be the projectionist. The basal ganglia decide which prefrontal behavioral movie to run and control the running of that movie.

Neural Output System
(How You Control Your Skeleton)

The final neural system we will examine is made up of the components we have just explored. We are going to utilize our analysis of those components to examine of how your skeletal control system functions. For this discussion of skeletal control, we are going to add one more component to our visual projection system analogy. The cerebral cortex is your storage projection screen, the thalamus is your lens-controlling projector, your projectionist is the basal ganglia, and the cerebellum is your movie-recording component. In this visual system there are two kinds of movies that are played. The first is procedural movies that are stored in the frontal cortex storage projection screen. The second type of movie is a recorded version of the first type stored in your movie recorder cerebellum. We will call this second cerebellar-contained movie a motor movie.

Skeletal muscle control is the first output behavior to evolve. Neural output to the skeletal muscles of your body is massively parallel and constant. Constant output is maintaining balance, providing muscle tone, supporting the body against gravity and doing mundane things like breathing. Your skeletal output system is organized hierarchically with the functions provided by newer neural components building on older systems. The older systems perform their duties unless overridden by neural input from above.

Many skeletal output routines do not involve body feedback. They are planned and executed explosively from start to finish. The act of throwing a ball to hit a target requires a plan that deals with a future time. The selection of a behavioral response takes into account the result of the behavior. The system is inherently forward looking.

We will begin with a re-examination of the various neural components that comprise your skeletal output system with an emphasis on functionality. Direct cortical input to motor neurons from M1 drives voluntary movement and motor neurons always fire in response to this input. All other input to motor neurons is provided by interneurons. Spinal cord and brain stem interneurons drive reflexes and rhythmic motion and integrate all of the separate non-voluntary motor systems that project down from above. The force of a muscle fiber contraction is directly proportional to the firing frequency of its motor neuron.

The inner portion of the spinal cord and brainstem are very old systems that control proximal muscles and perform non-voluntary duties such as balance and posture. The outer portions are relatively new and provide voluntary control of distal muscles. In humans, the corticospinal tract contains the axons of primary motor cortex (M1) pyramidal neurons driving this newer outer collection of motor neurons. The corticospinal tract first appears in primitive mammals and is utilized to modulate sensory input, not to control skeletal output. As we move up from primitive mammals to humans, sensory and motor cortex first overlap. Then distinct motor and sensory cortices appear with increased projections to spinal cord interneurons. Next, direct projections occur from motor cortex to outer spinal cord motor neurons and finally, direct motor cortex connections are also made to inner spinal cord motor neurons.

In primate mammals, corticospinal output provides the only direct control of distal limb movements. Monkeys have direct projections to the entire outer motor neuron pool and humans have extensive direct projections to both outer and inner motor neuron pools. Humans have a much higher degree of voluntary control over skeletal muscles than any other animal. The size of the primary motor cortex in all mammals is in direct proportion to body size. The size of the premotor frontal cortex in humans is six times the size of the same premotor area in monkeys.

The spinal cord and brain stem implement a variety of unconscious muscular controls projecting to inner pool interneurons. These systems control large muscle groups of the torso, shoulders and hips that are driven to affect a specific goal concerning the entire body. Anti gravity muscular output provided by the brain stem is an excellent example. Place your arm in front of you and rotate your hand. Neural output to motor neurons from your brain stem is counter acting the pull of gravity on your arm and hand as you move. You are not conscious of this function, because it is automatic.

Now pick up a heavy object and hold it out in front of you. Anti gravity muscular output is inhibited just prior to and during the voluntary movement associated with picking up the object. Once the object is held in front of you, your M1 firings to motor neurons only have to support the weight of the object. This output is added by interneurons to the anti gravity motor output of the brain stem and both your arm and the object are supported. The effort required to hold the heavy object seems conscious to you, because it is.

Spinal reflexes cannot be overridden by voluntary control. Brain stem reflexes can be overridden with various amounts of concentration required. If you have ever played the "fall backward and be caught by your friend" game as a child, you know that you can voluntarily override your brainstem control, but it is not easy and your limbic system does not like it

The cerebellum provides a learned modulating projection to each and every neural component that provides skeletal output. The output of cerebellar motor movies contains three distinct parts, balance, non-voluntary brain stem control, and voluntary movement. All behaviors that contain skeletal output are stored as a motor movie. If the pattern of cerebral input indicates that a learned motor movement is to be performed, the cerebellum will output the motor memory associated with that learned movement. Motor movies contain the actual firing sequences of primary motor cortex neurons utilized to drive learned movements.

The cerebellum is always learning what you are consciously doing with your body and adapting its recorded movie to match that conscious output. The inferior olives project climbing fibers to the neocerebellum that always cause purkinje neurons to fire and transition into learning mode. The climbing fibers of inferior olive neurons fire spontaneously every one to two seconds. This has the effect of stopping a small portion of the stored motor movie for a few frames. During these brief periods the affected neocerebellar purkinje neurons go into record mode and slightly modify their M1 output patterns to be more like procedural memory output.

This is how you learn all voluntary movement and adapt to a continuously changing body. There are small body changes literally every day. Muscles get stronger or weaker depending on use, you gain or lose weight and you age every day. Every motor movie stored in your cerebellum is altered continuously to support your body as it exists today.

What is most interesting about the neocerebellum and motor memories is how ubiquitous they are in driving your voluntary movement. Your neocerebellum participates in virtually every voluntary movement you make. When you perform any skeletal movement, like reaching for a cup, the neocerebellum recognizes the pattern associated with that task and outputs the motor movie to accomplish it. This allows you to attend to other things as you move about and physically interact with your environment. Most of the time your neocerebellum is driving your car.

The cerebellum is the easiest of your neural components to get in touch with. Simply monitor what your body is doing while you provide minimal conscious control. Perform any task while concentrating on something other than the task and I think you will be amazed at the level of skeletal control your cerebellum provides.

The thalamus performs exactly the same role for frontal cortex procedural memories as it performs for non-frontal declarative memories. The basal ganglia provide inhibitory control for every thalamocortical loop of the prefrontal, premotor and primary motor cortex. The basal ganglia control the selection of procedural memory and the premotor interpretation and primary motor implementation of that memory. One and only one output pattern at a time is allowed to drive skeletal muscle behavior.

Procedural memory patterns are created and retrieved in exactly the same manner as declarative memories. The cascade of synchronous pattern detection from prefrontal to premotor to primary motor cortex is exactly the same as the visual pattern recognition of an apple cascading through declarative memory areas with one exception, the basal ganglia enforce the one at a time rule.

Voluntary movement is a serial process and the temporal sequence involved in the generation of movement exhibits a wave of cortical activity. Cortical activity shows a readiness potential that develops just prior to actual planned movement. This movement readiness potential develops in parietal, temporal and especially in prefrontal cortex about 0.8 seconds prior to actual movement. The actual motor cortex fires approximately a twentieth of a second prior to muscle contraction. Between these two times the cerebellum becomes active and executes its role in output control.

Utilizing our visual system analogy lets examine what seems to be a relatively simple motor task, throwing a ball. You desire to throw a ball and hit a target. The situation in which you find yourself, the declarative and limbic input to the basal ganglia and frontal cortex, have resulted in your projectionist basal ganglia selecting a particular procedural movie called "throw a ball to hit a target". The declarative memory input includes a representation of space that contains the ball, the desired target and you.

The skeletal movement required to throw a ball has already been well learned. This means you have a motor movie of "throw a ball to hit a target" available in your cerebellum. In other words, the cerebellum has learned the sequence of pyramidal neuron output patterns in premotor and primary motor cortex that is required for the voluntary portion of throwing a ball. The prefrontal cortex stores the procedural movie built from all the past thrown balls.

Your prefrontal cortex sets up the variables required for this particular throw taking into account the weight of the ball, the distance to the target, the size and anticipated motion of the target, and the three dimensional space between the target and you. These variables result in premotor areas being primed to drive the skeletal muscle modifications required for this unique throw and selecting the amount of neural output that will be applied to various skeletal muscles to control the force of the throw.

All is ready. Fire the throw behavior movies. The brain stem nuclei that control housekeeping functions actually fire first. Balance and other non voluntary muscle activity associated with the throwing of a ball is part of the cerebellar learned ball-throwing movie. The basal ganglia projectionist now begins to run the procedural memory ball-throwing movie. The cerebellum-stored ball-throwing motor movie is automatically gated to the primary motor cortex through the thalamic gating neurons enabled by the basal ganglia. The motor movie from the cerebellum and procedural movie from the frontal cortex are synchronously played out. The cerebellum motor movie is utilized to control the throw unless the procedural movie overrides it with conscious control.

This results in a massive amount of parallel neural output to the entire body. The frontal cortex over rides portions of the cerebellar input to provide modification of angle and force. This modification action feels conscious to you as you throw, because it is. If you throw the same ball at the same target over the same distance enough times, the cerebellum will drive the entire action with no conscious intervention.

The movies end. The ball is thrown. Your basal ganglia projectionist now selects the "monitor the ball" behavior pattern and holds that movie active until the ball impacts. Based on success or failure to hit the target, changes to the variables are made to affect the next throw. Changes to procedural memories for these variables can be made with one experience, exactly like declarative memories. Changes to the stored ball throwing cerebellar motor movie are much slower and take repeated learning episodes.

Now a look at your complete voluntary skeletal movement output system. (Ref. Fig. 5-32 and Fig. 5-33) A motor neuron in your spinal cord exclusively innervates a group of your muscle cells. This motor neuron receives input on its 10,000 dendritic synapses directly from your frontal primary motor cortex (M1) and also receives input indirectly via spinal interneurons from all premotor areas and primary and secondary somatosensory areas that are concerned with the body location containing these muscle cells. This is a vastly parallel system.

These same primary motor, premotor, primary somatosensory, and secondary somatosensory cortex locations also project to the area on the pontine nuclei that provide input to the portion of the neocerebellum concerned with these muscle cells. That neocerebellar output is gated by the same gating type neuron in the thalamus that forms the thalamocortical loop with the primary motor cortex pyramidal neuron signaling to this group of muscle cells. These same primary motor, premotor, primary somatosensory, and secondary somatosensory cortex locations project to the area on the basal ganglia striatum that provides inhibitory control of this same thalamic gating neuron. The neural circuit we have just described controls the motor neuron output to just one group of muscle fibers. An identical neural circuit controls all of the motor neurons in your body. Combine all of these independent neural skeletal control circuits depicted in figure 5-32 together and you get the massively parallel system depicted in figure 5-33.

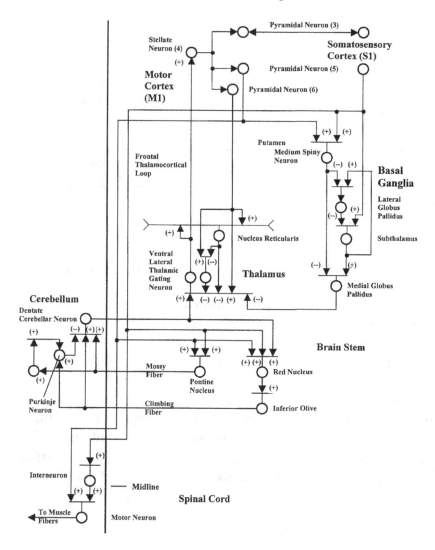

Figure 5-32: Motor Neuron – Voluntary Movement Neural Control Circuit

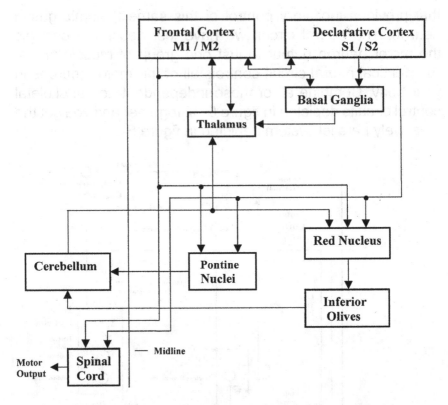

Figure 5-33: Voluntary Skeletal Movement

Both frontal cortex procedural memories and cerebellar motor memories are combined to effect voluntary skeletal movement. This is a continuous process. All of the systems and components described are continuously performing pattern detection and projecting their pattern recognitions onto the next component. Each motor neuron is controlled by a distinct set of neurons in the frontal cortex, basal ganglia, pontine nuclei, red nucleus, inferior olives, cerebellum and thalamus. Each of these parallel neural circuits is the same. They combine to produce the coordinated skeletal movement you take for granted.

Conclusion

Our analysis of the components that make up your brain is now complete. We have examined the available information generated by scientific research and utilized that information to apply analysis as to how these components function. These functional descriptions seem to be the most likely based on the currently available evidence. Next we will apply the same type of analysis to overall brain function.

Chapter 6

Brain Architecture Analysis
How Our Brain Works

Introduction

Each of your brain components has been presented and analyzed. Now it is time to combine those components together and examine your brain as a whole. We will begin with a review of the characteristics of neurons that have directly influenced our analysis. The three major human brain subsystems, the limbic system, the declarative memory system and the procedural memory system will also be reviewed. Finally, we will combine our three subsystems into a complete human brain architectural model and explore the capabilities of the human brain that interest us most. Functional descriptions of memory, learning, and human behavior will each be proposed.

Neurons - Pattern Detection / Synchronous Activity

The neuron is the sole active component utilized in the construction of the human brain. The neuron's dendritic structure supports synapses that convert external chemical input to internal electrical charge. The neuron's behavior is totally dependant on its internal electrical state completely derived from synaptic input. Neither the dendritic structure nor cell body store electrical charge. The axonal structure responds to an electrical charge threshold by firing and transmitting an action potential to all axonal synaptic locations and releasing the neurotransmitter it utilizes into the synaptic gaps. The neuron modifies its dendritic synaptic efficiency to become more sensitive to patterns that cause it to fire.

The Hebbian principle of synaptic modification is the only storage mechanism of pattern recognition. Synapses that are active when the neuron fires are strengthened. This is the storage mechanism of memory creation. Synapses that are active but do not cause the neuron to fire are weakened. This allows the pruning of synapses and enables the plastic characteristics of the cerebral cortex. All of the capabilities exhibited by the human brain are implemented by pattern detecting neurons. No other function will be attributed to neurons in building a model of how your brain works.

The neuron is a real time device that indicates its level of pattern recognition by the frequency of its output action potentials. The neural circuits that neurons build also respond to their current real time pattern input and produce action potentials that serve as input to other neural circuits. Every neural component in your body is a real time entity. Neural input at every level is continuous and constant. Each successive neuron, neural circuit, and neural component is continuously responding to pattern input and producing pattern output that drives the next level of neural function. Your brain is a real time device, responding to the current patterns of input and producing patterns that drive your behavior. The brain stores nothing.

We have postulated that neurons form two distinctly different types of neural structures, genetically controlled hardwired structures and homogenous neural arrays that are wired with learning. Neurons that comprise hardwired neural circuits develop and prune their synaptic connections in the same manner as neurons forming learning type circuits. The patterns available to these neurons support a neural function that is fixed and does not change. Therefore the pattern recognitions and neural circuits that these neurons form are static or hardwired. The complexity and diversity of the evolutionary tuned hardwired portions of your brain are impressive but we are now concerned with the portions of your brain that enable learning.

The neurons that support learning are the purkinje neurons of the cerebellum, the medium spiny neurons of the basal ganglia and the pyramidal neurons of the cerebral cortex. These neurons all support extremely large numbers of synaptic inputs. The average number of dendritic synapses for a cortical pyramidal neuron is 60,000, for a striatal medium spiny neuron 10,000, and for a purkinje neuron 200,000. The evolutionary growth of these dendritic arbors to receive this number of synapses argues strongly for a neuron that detects some basic correlation between vast numbers of inputs. The correlation between synaptic inputs that these neurons detect is synchronous activity. The utility of detecting patterns of inputs firing synchronously is enhanced by larger numbers of inputs. The greater the number of inputs to an individual neuron, the greater the synchronous pattern complexity detected by a single neuron.

A memory is the recognition of a previously encountered pattern. The actual modification to an individual synapse on the dendritic spine of a firing pyramidal neuron is very small. A pyramidal neuron that fires due to receiving a particular synchronous pattern on its dendritic tree will slightly strengthen all of the synapses that contributed to that firing. The same synchronous pattern is now ever so slightly more likely to cause this neuron to fire. Perhaps more importantly, a somewhat degraded version of the same synchronous pattern is now slightly more likely to cause this neuron to fire.

A memory pattern involves an area of cortex and is supported by a very large number of pyramidal neurons. The number of neurons in a square millimeter of cortex in mammals is around 150,000. The percentage of pyramidal neurons in non-granular cortex of rats is approximately 92%.

Those approximations yield a figure of 138,000 pyramidal neurons in one square millimeter of association cerebral cortex. One square inch of your cortex contains in excess of 89 million pyramidal neurons. An average of 60,000 synapses per pyramidal neuron yields the incredibly large number of 5,340,600,000,000 pyramidal synapses in one square inch of your association cortex. Extremely small individual synaptic enhancements over literally billions of synapses add up to represent a large overall change. Billions of extremely small synaptic enhancements due to the recognition of a synchronous pattern over an area of cortex are what enables your pattern recognition memory capabilities.

The Human Limbic System (Nature vs. Nurture)

Most of us believe that we are in control of our emotions most of the time. Your emotional response generating limbic system is very old, highly evolved and, with the exception of the learning capable limbic cortex, genetically inherited from your ancestral tree. The cerebral neocortex, thalamus and basal ganglia are the new kids on the block and in many ways the older limbic system remains very much in charge. Unless you exert conscious control, it is in charge. As with all evolutionary changes to the brain, the older system has not been replaced. It has been enhanced with learning ability and conscious oversight.

The limbic system consists of hardwired limbic nuclei and the archeocortical limbic lobe. (Ref. Fig. 6-1) The original role of your limbic system is the control of the hormonal state of your body. The hypothalamus provides this control and is made up of a large collection of highly interconnected nuclei with various output responsibilities. The hypothalamus is the central command output center of your limbic system.

Add limbic cortex to the hardwired nuclei of the limbic system plus a small amount of additional cortex and you have the equivalent of a crocodile's brain. Lets examine how the limbic system controls a crocodile's behavior. A crocodile passes a mound of mud that happens to be a turtle's nest full

239

of eggs and detects the smell of a dead egg in the nest. The crocodile's limbic system responds to the smell of something rotting by driving the "find what is causing this smell and eat it if possible" hardwired behavior. The crocodile digs into the nest, discovers the eggs and eats them.

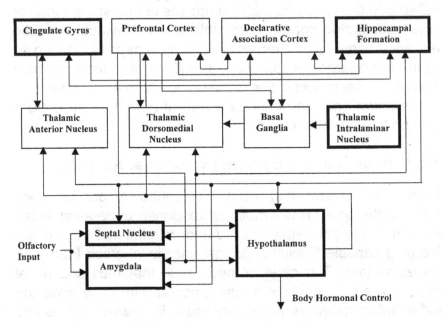

Figure 6-1: Limbic System

A few days later, the crocodile comes across another turtle nest mound of mud. Sensory input concerning the mound of mud activates the object pattern for the mound of mud stored in the crocodile's first encounter. The activation of this object pattern causes the crocodile's hippocampus to generate the emotional state associated with that object and that emotional state is associated with eating. The limbic response to this nest object makes the crocodile want to eat the nest, it looks like a piece of your favorite pie. The crocodile has no clue what is inside the mound of mud. Associating the mound of mud object with the object pattern of turtle eggs is way beyond the crocodile's capability. The result is the same. The crocodile digs into the nest and eats the eggs.

A human observing the crocodile would assume that it knew that the nest contained eggs. The crocodile knows nothing. Its brain does not contain enough cortex to build associations between objects. It deals with the world one simple object at a time. Its limbic system does generate the emotional state tied to those simple objects and causes the crocodile to behave towards those objects with hardwired behaviors that are appropriate. Your limbic system does for you what the crocodile's does for it. The hippocampus and limbic system pattern detect objects and behavior and evoke the emotions associated with them.

Your limbic system is as hardwired as a crocodiles but it is not identical. The human evolutionary path separated from reptiles long ago and the human limbic system continues to evolve. An example of this is worry. Humans worry and crocodiles do not. Declarative objects that humans tend to worry about are their children. It is intuitive that the human children of parents that worried about them would have a survival advantage over the children of parents that did not worry. As a result of that survival advantage, when humans think about their kids they have a tendency to be concerned, to worry. In other words, when the declarative objects that represent their offspring light up in the declarative memory of humans their limbic systems produce an emotional response of protectiveness and concern that we call worry. The hypothalamus of humans has evolved to produce this emotional response of worry.

Your limbic system provides input to two higher level neural processes. Limbic input is crucial to the storage of new memory patterns, both declarative and procedural. This memory write function is provided by the hippocampal formation that includes the hippocampus proper and surrounding supporting cortical areas. The hippocampal formation is reciprocally connected with all associative areas of cortex to facilitate its memory write function.

The second neural process that utilizes limbic input is the selection of procedural memory. Limbic input is combined with declarative input in the prefrontal cortex where procedural

241

memory patterns are recognized. The entire prefrontal cortex is primary sensory input cortex for limbic input. These interconnections are not localized as in other primary sensory cortex but are spread diffusely throughout the prefrontal cortex. Limbic sensory input to the prefrontal cortex comes mainly from the hypothalamus and the amygdala that represent the main and last stages of the limbic emotional system. Limbic input from the intralaminar nucleus of the thalamus to the basal ganglia determines the level of dopamine production and thereby influences which procedural memories are chosen for execution. You demonstrate quite different behaviors to a particular situation depending upon how you feel.

Prefrontal cortex allows you to become conscious of your emotional state. Prefrontal cortex via its output projections to the amygdala and hypothalamus has the ability to strongly influence and control the emotional state provided by the limbic system. You can do much more than make yourself happy or sad by thinking of actual or fictitious situations that elicit those emotions. Humans can suppress fear and ignore hunger or thirst. Humans can override limbic input with conscious control.

The human limbic system has evolved to include the conscious influence of the prefrontal cortex as an intimate partner in determining mood, emotions, and motivation. Your limbic system drives your behavior unless you exert conscious control. The continuing argument of nature verses nurture is about old limbic system verses prefrontal cerebral control over behavior.

Your Declarative Memory System

Declarative memories are what we usually think of as memory, the recall of people, places, and things. Your declarative memory system is implemented by three major components, the non-frontal cerebral cortex, the thalamus, and the hippocampal formation. The fundamental neural functionality that allows this system to operate is the cascaded detection of synchronous patterns of input by cortical pyramidal neurons under the timing control of the thalamus.

You have stored an amazingly large repertoire of declarative memory patterns for both objects and space within your cortex. Sensory input patterns from objects and space not previously encountered are recognized by the pattern recognition they elicit from previously stored patterns. You evaluate your world exclusively by comparing it with your stored representation of it.

We are again going to employ the movie projection analogy utilized in our previous discussions. The cerebral cortex is the storage projection screen, the thalamus is the projector and the hippocampus causes images to be stored within the screen. This movie is real time. The thalamus projector does not shine a steady light onto the screen. It sends a continuous stream of pulsed energy that allows the screen to light up and keeps images on the screen in resonance so they can be recognized. The cerebral cortex projection screen is not an ordinary screen. The images it lights up are internally generated. Real time sensory input compares with all previously recognized images and the images that most closely compare light up on the screen. The brightness of an image is proportional to how well the new image compares with it. New images are continually being stored and existing images modified with new input.

For a declarative memory to be accessed it must first be stored. Each and every active synapse is slightly enhanced when a declarative memory achieves dominance. The details of this synaptic modification process are unclear but we can provide a high level description. The hippocampal write signal causes a temporary modification in active synapses and triggers subsequent protein structural changes that make the change permanent. If the structural change is prohibited, the temporary modification is lost within hours. The temporary modification and permanent replacement are equivalent in their ability to allow the new pattern to be recognized.

The hippocampus causes the storage strength of the current image to be proportional to the limbic system's response to the current image. Important images are stored with a great deal of strength and will light up easily when new movie input compares with them. Non-important new images will be stored weakly and unless recalled quickly with new input may never light up the projection screen and be lost.

The thalamus projector provides the management of thalamocortical loops that enables the declarative memory system to operate. The thalamus controls the timing window within which incoming patterns are sampled by cortical areas, disables all but the strongest signaling thalamocortical loops, and also binds cortical areas that pattern recognize strongly into a synchronous collection of oscillating thalamocortical loops. All pattern recognition flow subsequent to primary sensory input is supported by reciprocal connections between cortical areas. It is through the control of thalamocortical loops that pattern recognition flow is built into declarative memory.

Vertical collections of cortical pyramidal neurons act as a whole in implementing memory. In primary sensory cortex these vertical collections of neurons form distinct columns that each process a specific characteristic of incoming sensory input. Although these column organizations are much less distinct in association cortex, association cortex functions as an organized vertical collection of neurons analogous to primary sensory cortex. Association cortex does not exhibit the fixed area usage of primary sensory cortex.

Each cortical area operates in relative isolation. A single cortical area makes up a small portion of a memory. It is the cascaded pattern recognition of successive cortical areas, utilizing a standard identical cortical area function, that forms a whole memory. A declarative memory is represented within the cortical sheet as a unique synchronous pattern of signaling cortical areas that occupy the entire cascaded chain of declarative memory cortex.

Your storage projection screen contains all of the declarative patterns that define you and your world. Your storage projection screen is not a layered storage container; it is literally one image deep. Every one of your declarative memories is completely intermixed within this one layer. Your entire life is stored as an intermixed set of patterns. An input pattern self selects the closest matching pattern and causes that pattern to appear out of the myriad of patterns stored. The entire screen is continuously displaying the closest matching stored patterns.

If you could actually look at this projection screen, the various patterns and their intensity would convey the state of declarative pattern recognition. Cerebral projections to non-cortical neural components are an exact reflection of this overall array pattern and the pattern projected to each non-cortical component is exactly the same. That projected declarative pattern is also completely synchronous, which allows it to be pattern recognized by non-cortical components.

There are many patterns active within the declarative cortex at any one waking moment as continuous memory forming waves wash continually through the declarative cortex. A particular cortical area may be a part of multiple declarative recognitions at the same time. In a conversation with four friends, each of their faces has a unique overall cortical pattern but each of those patterns share large portions of the same cortical screen. Each unique facial pattern gains dominance as you saccade from face to face.

The particular pattern that represents a declarative memory changes with use. New input patterns that elicit memory are stored and modify the current memory pattern. Every access of a previously stored declarative memory alters the stored pattern to incorporate the new pattern that caused recognition. With continued recognition, a declarative memory pattern becomes more efficient. It actually shrinks in area size and signals more strongly when activated. Its center signals pattern recognition more strongly with use and the edges of the pattern are inhibited and lost.

The input utilized in memory recognition comes from forward propagating cortical input, hippocampal input, contralateral cortical input, and cortical feedback from higher association areas. (Ref. Fig. 6-2) Forward cortical input is the primary pattern source for memory recognition concerning the external world. Hippocampal input determines the intensity with which the current pattern will be stored and cortical contralateral input serves to keep the two brain hemispheres in the same state of pattern recognition.

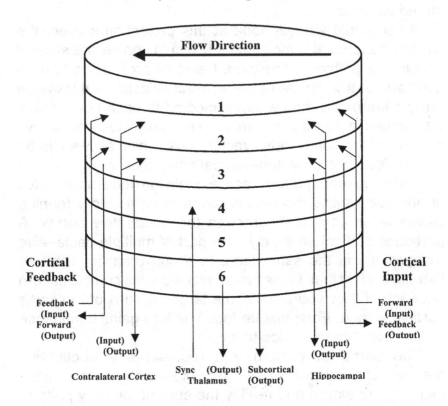

Figure 6-2: Cortical Flow

Cortical feedback serves a number of functions. In support of memory formation it indicates whether the cortical output remains part of a forming memory. This allows the final pattern image to be sharpened to only include relevant areas. Cortical feedback also plays a critical role in maintaining background declarative objects active within your storage projection screen. When you enter a room that contains a chair, the declarative pattern for the chair is projected to the parietal cortex and becomes part of your virtual environment. Feedback from the parietal to temporal cortex keeps the chair object pattern active within your declarative storage projection screen even when visual input of the chair object is lost. If you leave this space, your virtual world no longer includes the chair, the parietal chair feedback is lost, and the object pattern turns off.

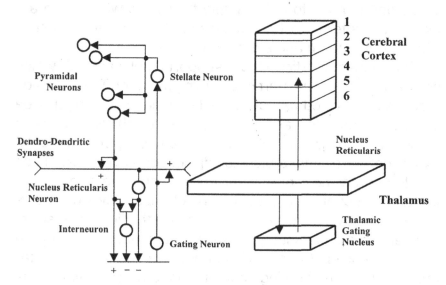

Figure 6-3: Thalamocortical Loops

The interconnectivity and functionality of thalamocortical loops provides the core capabilities that enable your declarative memory system. (Ref. Fig. 6-3) Understanding the functions that thalamocortical loops provide and how they provide those functions is a prerequisite to understanding how memory works. Each area of cortex is supported by thalamocortical loops and those loops must be active in order for the cortical area to be active. The two dimensional pattern of active declarative memory is exactly reflected in the two dimensional pattern of active loops enabling that pattern.

The thalamus has complete control of the functions supported by thalamocortical loops. Gating neurons drive each and every thalamocortical loop. The nucleus reticularis dendritic structure is completely interconnected by dendro-dendritic synapses, collateral inputs are received from both projections of every thalamocortical loop, and inhibitory control of every thalamic gating neuron is provided. This allows the

247

thalamic nucleus reticularis to provide integrated control of the entire cerebral cortex.

The thalamocortical loop provides a timing window that gates input to cerebral pyramidal neurons. This function causes pyramidal neurons to only sample their input when there are legitimate patterns to be recognized. This has the effect of filtering out random and extraneous noise.

The thalamocortical loop supports selection of which cortical areas become part of a forming memory. This selection process is a shared responsibility between local cortical area inhibitory processes and the larger scale inhibitory nucleus reticularis. Cortical pyramidal neurons support horizontal axonal collaterals that project to adjacent cortical areas. These collaterals are most prominent from output layers three and five and interconnect the entire cortex horizontally. This allows inhibitory stellate basket neurons to detect the firing intensity of adjacent areas and shut down their area if pattern recognition in other areas is stronger. This provides a fine focus of detail resolution to the forming memory pattern. This allows the nucleus reticularis to only provide inhibitory control for larger scale areas of memory formation.

The selection of cortical areas that remain part of a forming memory is also a function provided by thalamocortical loops. Each cortical area that becomes part of a forming memory pattern must receive positive reciprocal feedback in order to remain active. Cortical areas not receiving positive feedback are inhibited. This serves to sharpen the final declarative pattern.

The final functionality provided by thalamocortical loops is synchronization. Areas of cortex become synchronized because their supporting thalamocortical loops are caused to be synchronous. This is accomplished by causing all of the gating neurons that drive the thalamocortical loops to fire at the same time at the same frequency. This supports the continuous synchronization of a memory pattern as it grows. This same capability provided by the nucleus reticularis allows widely separated areas of cortex to be synchronized.

We will now examine memory formation in more detail. The recognition of a declarative memory forms as a wave of synchronous input roles across the declarative cortex causing cortical areas that become part of the forming memory to become synchronous with the forming memory. The nucleus reticularis detects incoming active thalamocortical loops and allows gating neurons to fire and provide the first synch window. That synch signal is distributed by the stellate neurons to the dendritic arbors of all pyramidal neurons in the area and removes all magnesium blockages. The forward cerebral input to the cortical area is then sampled. If the excitatory input synchronous with this sampling window is large enough, the pyramidal neurons of the area all signal with strength proportional to their level of pattern recognition.

Stellate basket neurons in the cortical area shut down pyramidal neurons if adjacent areas are projecting more strongly. Output from layer six signals the pattern recognition strength to both the reticularis and gating neurons of the thalamus. The nucleus reticularis inhibits the gating neurons if the pattern recognition strength is weaker than other areas. Gating neurons that are inhibited turn off and their supported thalamocortical loop is broken. This focuses the declarative memory to only the strongest comparing cortical areas.

As this forward propagating wave of declarative memory formation washes through the cortex, each successive area of pattern detecting cortex sends a backward ripple that indicates which areas remain part of a forming declarative memory. Areas that receive positive feedback increase their signal strength and this allows the nucleus reticularis to shut down the thalamocortical loops of areas not receiving a positive ripple feedback. The end result of this forward propagation wave of pattern recognition and ripple feedback is that only the strongest signaling areas that comprise the winning declarative memory remain on and synchronous within the cortical projection screen.

Reticularis synchronization of gating neurons causes cortical pyramidal neurons to synch up with the incoming cortical pattern that caused them to fire. The firing frequencies of the thalamocortical loops that are added to the forming memory become synchronous with the thalamocortical loops of the partially recognized memory. The newly enlarged memory recognition pattern is then passed to the next areas of association cortex and the wave propagates through the entire declarative cortex. The thalamocortical loops of widely disparate areas of cortex thus become synchronized and form a coherent, whole declarative memory.

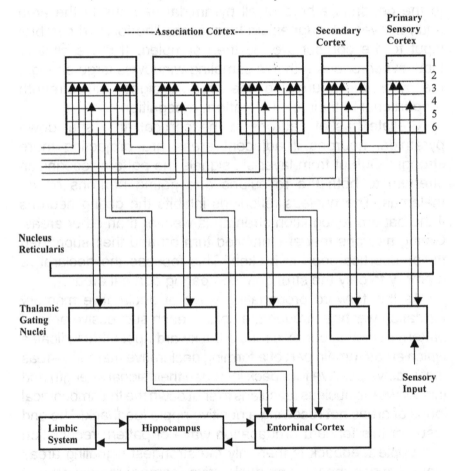

Figure 6-4: Declarative Memory System

Lets explore the recognition of an actual declarative memory. The recognition of a memory from cortical input starts in primary sensory cortex. (Ref. Fig. 6-4) We are going to use visual input for our example but all modalities of sensory input, including olfactory, can elicit memory recollection. Light falling on the electrical gap interconnected retina causes a synchronous static two-dimensional array to be projected through the thalamus to the primary visual cortex (V1). Columns of V1 that receive patterns that are consistent with their training, signal that pattern recognition to all target locations. Secondary visual cortex areas sensitive to more complex visual features are enabled to sample the output of the primary visual cortex. If during that thalamic controlled window they receive a large enough synchronous input that matches the pattern they have learned to recognize, they will signal to all output targets with a strength proportional to their level of pattern recognition.

Each successive area of secondary visual cortex has learned to recognize a more complex visual pattern over a larger visual area. The enabling window provided by the thalamus proceeds like a wave across the cortex. From primary to secondary and on to successive layers of association cortex this enabling wave allows cortical areas that pattern recognize the synchronous input to signal that recognition and become part of the forming memory. Cortical areas that do not signal strongly or signal less strongly than another area are inhibited. Strongly signaling cortical areas are continually synchronized by the nucleus reticularis to form a coherent synchronous pattern available for recognition by the next cortical area. A coherent pattern is constructed within the cortical sheet as the enabling wave passes. A cortical area that is a part of the winning pattern signals in both a forward direction to propagate the memory formation and also in the reverse direction to reinforce the winning input. Cortical areas that do not receive positive feedback are dropped from the forming declarative memory.

At some point in this wave propagated memory formation we have visual object recognition within the temporal lobe. Lets assume it is your grandmother's face. Visual object recognition of your grandmother's face is represented within the cortex as a cortical pattern of firing, synchronized vertical areas that encompass the occipital cortex and a good deal of the temporal cortex. The visual object recognition of your grandmother's face occupies a large amount of declarative cortex. The enabling wave that got us to this point does not stop here but continues through the entire declarative cortex to tie together your grandmother's name and all of the other associations triggered by the memory of her face.

The cortical pattern on your cerebral projection screen that represents your grandmother contains cortical patterns that represent generic facial objects such as noses, eyes, ears, etc. The concatenation of those visual objects combines to build a unique cortical pattern. Your declarative cortical projection screen is continually lit up with a variety of object memories. A particular area of cortex is a part of many memories. A cortical area that recognizes the visual pattern associated with noses is a part of larger cortical patterns that represents faces, humans, mammals, primates, and your grandmother.

When the declarative memory forming wave reaches the highest levels of declarative association cortex, areas in the parietal and temporal cortex, all object and spatial memory patterns recognized in the current primary input are on in the declarative cortical projection screen. These synchronous patterns are then passed forward to the procedural memory system in the frontal cortex.

Your Procedural Memory System

Most of your procedural memory system is an exact replica of your declarative memory system. The procedural memory system includes an additional two components, the basal ganglia and the cerebellum. The basal ganglia is required because your procedural memory system defines and drives your behavior and only one behavioral controlling procedural memory can be allowed to be active at one time.

The cerebellar system alleviates the problems associated with the complexity and continuous amount of pattern output required for control of skeletal behavior. There are two separate and distinct memory systems involved in skeletal behavior. Procedural memory in prefrontal cortex defines overall behavior and motor memory residing in the cerebellum provides learned motor output. Skeletal muscle control and voluntary movement are supported by a combination of these two memory systems.

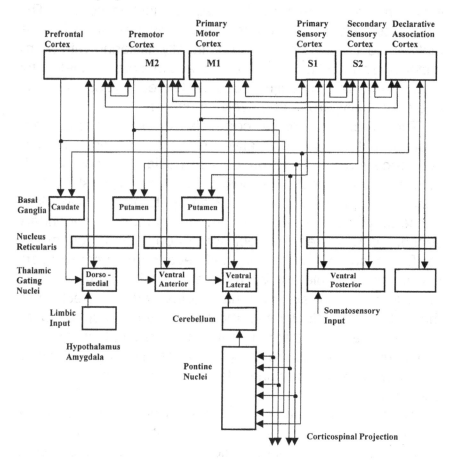

Figure 6-5: Procedural Memory System

Three procedural memory system components are identical to the components of the declarative memory system. (Ref. Fig. 6-5) The cerebral cortex, thalamus, and hippocampal system perform exactly the same roles in the procedural system that they perform in the declarative system. There is no difference in the neural makeup or functionality provided by these components between these two systems.

The frontal cortex is a one image thick storage projection screen that stores all of your learned behaviors as an intermixed set of patterns. An input pattern of declarative and limbic input self selects the closest matching behavioral pattern and causes it to light up in your frontal storage projection screen. The difference between your procedural and declarative screens is that the declarative screen is continually flickering with many patterns representing objects and space that are competing for recognition while the entire procedural screen is occupied with one continually strong and evolving pattern.

The procedural frontal cortex is completely reciprocally interconnected with declarative cortex. Each frontal cortical area receives from and projects to declarative cortex. This massive interconnection is most dense in the highest levels of declarative and prefrontal cortex and gets progressively less dense as you progress away from these areas. All secondary sensory cortex areas project some connections forward to the frontal cortex and primary and secondary somatosensory cortex are completely interconnected with premotor and primary motor cortex. Prefrontal cortex pattern detects both the state of your declarative memory system, people, places and things, and the state of your limbic system, how you feel. Procedural memories are recognized based on how close your virtual world and emotional state compares with previously encountered situations.

The thalamus is again your projector but in the procedural memory system you have a projectionist, the basal ganglia. Each thalamic lens must be allowed to perform its standard function by the basal ganglia. There is no large wave of memory formation that washes over the procedural cortex. Procedural memory flows from one pattern to the next under basal ganglia control as one strong pattern continues to evolve.

When not disabled by the basal ganglia, each thalamic lens provides the same functions of pyramidal input gating, memory pattern focus, and synchronization. The synchronization provided here causes the procedural memory pattern to become synchronous with the declarative input pattern that enabled it. The thalamus supports this synchronization of widely separated memory patterns by causing the gating neurons and thalamocortical loops that support those patterns to be synchronous.

The hippocampus again drives the storage strength of the current procedural image to be proportional to the limbic system's current emotional state. Behavior is modified exactly as declarative memory is modified. Often employed behavior patterns shrink in area and signal more strongly when recognized. Only one behavioral pattern at a time dominates the procedural screen and a selected behavior tends to occupy the screen for some time. Behavioral modification is a relatively slow process.

The projectionist basal ganglia control the flow of procedural memory that defines behavior. This projectionist takes as input all that is active on the entire projection screen and based on that input, filters the image on the procedural screen to one sharp, well defined series of image patterns. That pattern flow is held by the basal ganglia until the behavior caused is complete. The basal ganglia's individual thalamocortical loop controllers are bi-stable devices that are normally disabling but can easily be maintained in an enabled state once activated. This allows a procedural memory to be maintained in an active state by the procedural memory input alone, even after the enabling declarative memory pattern has disappeared.

The motor movie record and playback cerebellum provides the second memory component of your procedural system. The cerebellum stores skeletal motor patterns and the cerebral state associated with them. The cerebellum is a replay machine that projects onto the skeletal output portion of your cerebral screen through the thalamic projector. The cerebellar system is watching the whole ongoing movie on your cerebral projection screen at all times and storing segments of movie that contain skeletal output. If a portion of the movie is familiar, the cerebellum plays back the skeletal output portion. This playback cerebellar system is a completely separate and distinct memory system. The input that is utilized for the selection of replay segments by the cerebellum is identical to the input utilized by the basal ganglia in the control of procedural memories.

The cerebellum is always adapting to changes in skeletal control, even when it is driving motor output. Each individual controller of an output pixel of motor movies is caused to enter a learn state every second or two. This causes the cerebellum to alter that pixel to incorporate the latest voluntary output. The motor movies stored in your cerebellum are always a reflection of your current physical state.

Using our movie projection system analogy and figure 6-5, we will now follow the flow from one frame of your procedural movie to the next. First, we freeze the movie at one frame. The current declarative memory pattern defining your virtual world is lit up and stable at the basal ganglia, the nucleus reticularis and the cerebellum. The current procedural memory in your frontal cortex driving your current behavior is also lit up and stable at the basal ganglia, the nucleus reticularis and the cerebellum. These procedural and declarative memories have been synchronized by the nucleus reticularis. The neurons of the basal ganglia that are enabling the current procedural memory are in an up state. The cerebellum is pattern detecting this movie frame and if it contains skeletal output, projecting the learned pattern of skeletal muscle contractions associated with it.

Now lets role the movie forward one frame. The wave of declarative pattern recognition has propagated to the highest levels of declarative association cortex and the resulting pattern in these areas project forward to prefrontal cortex. Lets assume that this new wave of declarative memory formation contains new input about your virtual world pattern detected from sensory input. This wave is not going to be allowed to wash over the entire frontal cortex. It is going to be restricted by the basal ganglia.

This new declarative input to the basal ganglia causes the neurons that receive it to enter their up state. The gating neurons in the thalamus that these neurons control are therefore not inhibited. This allows the new declarative input to be pattern detected by prefrontal cortex that receives it as forward input. The nucleus reticularis detects this incoming wave and allows a sampling window to be projected to the prefrontal cortex. Prefrontal cortical areas that receive this sampling window and detect a large synchronous input, signal with strength proportional to their degree of pattern recognition. All of these cortical areas project toward premotor cortex, back to the declarative cortex that inputted to it, and to the basal ganglia, nucleus reticularis and cerebellum.

The procedural memory pattern is modified due to the new declarative input. Cortical stellate basket neurons and the nucleus reticularis suppress weaker signaling areas to limit the resulting pattern to only the strongest signaling areas. The basal ganglia select the actual procedural memory pattern. Only one procedural pattern will remain enabled and be allowed to drive behavior. The feedback of this final selected procedural memory to declarative association cortex will cause all declarative areas not receiving positive feedback to be inhibited. The current procedural memory is probably the largest factor in determining which areas of prefrontal cortex remain enabled. Your behavior does not shift radically unless your declarative cortex detects something extraordinary.

The restricted wave of pattern detection in the prefrontal cortex now propagates to M2 and onto M1. The active prefrontal area pattern is pattern detected by premotor and primary motor cortex to effect the skeletal output portion of the current procedural memory. This cortical flow is controlled in the same manner as prefrontal cortex. The enabling sampling wave flows from enabled prefrontal areas to premotor and motor cortex. The actual motor output cortex is also receiving any learned motor memory from the cerebellum that may be relevant to the indicated behavior. The prefrontal pattern will determine if the cerebellar input is to be used. The modification of prefrontal patterns with learning to utilize motor memory directly is a part of the learning process for skilled skeletal movement.

Prefrontal areas that remain enabled are now synchronous with the declarative input that enabled them and form one large synchronous memory pattern. You now have a new frame of the movie of your life.

This process is real time and ongoing. Waves of new declarative input arriving at the frontal cortex, restriction of that wave to only areas receiving significant synchronous declarative input, pattern detection of that declarative input in conjunction with limbic input in the prefrontal cortex, selection of one procedural memory by the basal ganglia, the forward propagation of the restricted wave to areas M2 and M1, and the synchronization of the final procedural memory with the declarative memory input that caused it. Procedural patterns flow smoothly from one pattern to the next as declarative and limbic input changes your current behavior.

Brain Architecture

The architecture of the human brain is realized by joining the declarative and procedural systems into an integrated whole. (Ref. Fig. 6-6) The only interconnection between these two systems is the reciprocal connections between the cortical areas of both systems. All thalamic, basal ganglia, contralateral cortical connections and hippocampal connections are strictly segregated between the two systems. The only communication path between the declarative patterns of your virtual world and the procedural patterns driving your behavior is the reciprocal interconnections between pyramidal layers two and three of the declarative and procedural cortical sheets.

Your cerebral cortical sheet is physically one continuous entity. Logically it can be viewed as two separate and distinct projection screens that are completely interconnected. The current state of your cortex defines your virtual world and your response to it. This current all encompassing memory pattern is created and maintained by waves of synchronous memory formation. Declarative images that cause the winning procedural memory receive reciprocal feedback from the procedural screen that maintains them as part of the forming whole memory. The intensity of declarative images not receiving positive procedural feedback is diminished in comparison to those that do.

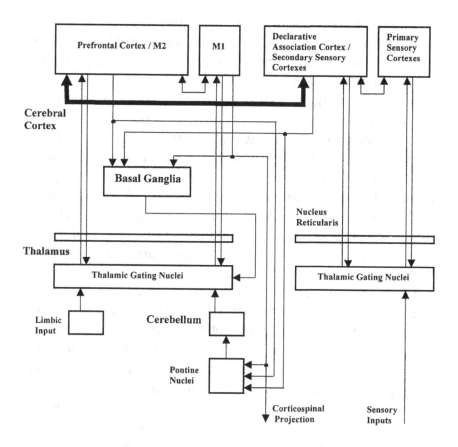

Figure 6-6: Brain Architecture

The final winning whole memory occupies the total portions of both screens. The majority of those screens are a mirror image of one another. Each and every declarative pattern that is a part of the whole memory is receiving procedural positive feedback. The procedural memory pattern driving your behavior contains a mirror reflection of the specific declarative input pattern that enabled it.

The hardwired thalamus provides the control that enables the formation of a whole cortical memory. That control facilitates the recognition and synchronization of memory within the cerebral projection screen as a whole. The timing of thalamic gating neuron firings is the key to both memory formation and synchronization across the entire cortex. The thalamus causes the gating neurons that support an entire memory across the entire cortex to fire in unison. The entire neural infrastructure supporting the current complete memory pattern is caused by the thalamus to be in resonance.

The basal ganglia is not involved in either the timing of memory formation or the synchronization of memory formation. The basal ganglia select the winning procedural movie and hold that movie active until the behavior caused is accomplished. The cerebellar system is a separate and distinct entity that sits outside and monitors the entire cerebral movie screen. It allows learned skeletal control to drive physical behavior without conscious prefrontal intervention. The majority of skeletal output always comes from the cerebellum, even for movements controlled consciously.

It is the interaction between the declarative and procedural projection storage screens that defines what you perceive and drives what you do. Declarative patterns that receive procedural positive feedback have your attention. A declarative memory gains your attention when the procedural mirror pattern image lights up in response to it. This is what defines attention. The declarative patterns of objects and space that are enabled by procedural feedback and therefore have your attention are in resonance with the winning procedural memory.

Procedural memory both enables and drives declarative memory. The key to higher-level brain function in humans is the procedural to declarative cortex feedback path. Procedural memory patterns have the ability to cause declarative patterns to become active that have no synchronous forward input. The reciprocal feedback path from procedural memory can elicit declarative memory. The capability of a procedural memory pattern to activate specific declarative memory patterns that are then synchronized with it by the nucleus reticularis is fundamental to higher-level human brain function. Procedural memory has the ability to cause declarative memory patterns to become active that are not generated from external input. This is what we humans term thinking.

If not for basal ganglia control of the frontal procedural pattern, the mirrored images of your declarative and procedural storage screens would each always reflect the other exactly. The procedural screen reflects the declarative if allowed by the basal ganglia. The declarative screen always reflects the procedural. A procedural memory pattern in prefrontal cortex that drives declarative memory recognition contains the reciprocal feedback pattern for that declarative memory.

Declarative memories caused to become active because of procedural projections from the frontal cortex propagate the entire declarative pattern that can cause the declarative pattern to become active. When you think of the word apple, the entire declarative pattern for apple, including everything you know about apples, lights up in your declarative screen.

The declarative patterns that are caused to become active solely by the procedural system always have your attention. Procedural control of the declarative system allows you to completely ignore all input from the environment and drive your procedural memory system from declarative memories you choose. This ability gives you complete control over the declarative input you choose to be driven by. This is imagination, daydreaming, creativity, thinking, all of the mental attributes we collectively call human.

The synchronization of the entire current active cortical area pattern is the key to overall brain operation. The synchronous projections of the current declarative and procedural active cortical areas enable pattern recognition in the hippocampal formation, the basal ganglia and cerebellum. This enables the selection of the next procedural memory and motor memory that accompanies it. Our architecture model of the human brain is now complete. The next step in analyzing how your brain functions is to use this model to examine how your brain actually works.

Declarative Learning – Storing the Patterns of Your Virtual World

Humans have the ability to recognize an object through pattern recognition of any primary sensory input. If you see an apple, you immediately recognize the visual patterns associated with an apple. It does not matter what the color or size of the apple is or at what angle you view it, you have modified your apple object pattern to include every visual input pattern that has ever evoked it. If you put on a blindfold, you can recognize an apple by its shape and texture by holding it. An apple's taste and smell will also easily identify an apple. Any one of your primary senses provides sufficient information to recognize an apple.

What is unique about the human brain is its ability to recognize objects through pattern recognition of abstractions of those objects in the form of words. The spoken and written word apple are just as effective as primary sensory apple input in lighting up the apple object in your declarative cortex. Human associations with the object apple do not stop there. An apple is a fruit that grows on trees. You know the names of several varieties of apples and perhaps where they are grown. You are familiar with apple pie and associate it with various holidays or people. Apple cider, apple strudel, the list goes on and on. Contemplate an apple long enough and you can build an impressive array of associated memories and connections with other objects. What you can do with the apple object, you can do with any object. Understanding how this complex web of connections and abstractions is constructed within your brain is crucial to understanding how humans build the declarative patterns that enable their intelligence.

Our quest for that understanding will begin with an examination of how the association of objects and the words that represent them are constructed. For two declarative objects to become associated, an association pattern object that ties the two objects together must be built into declarative association cortex. When two unrelated objects are caused to be synchronous, their forward synchronous projections will propagate as waves through higher-level association cortex.

262

A pattern of association cortex will fire in recognition of where their waves are additive. That association pattern ties the two objects together. The synchronization that causes two previously unrelated objects to become associated requires interaction between the declarative and procedural systems.

When you encounter a new object and its declarative object pattern achieves dominance, the cortical areas that represent the object project forward to your frontal cortex. The procedural cortex receiving this forward projection lights up and projects a mirror image feedback to the new declarative object pattern. That object has your attention and the mirror images of the declarative and procedural object patterns are in resonance.

We will assume at this point that someone tells you the name of this new object. Hearing the name causes an audio word object to light up in your declarative projection screen and project forward to your frontal cortex. The current procedural memory pattern is allowed by the basal ganglia to receive and include the audio word object and its mirror image is projected back to the declarative screen. At this point, both the new object and audio word object have your attention and are synchronous with the current procedural pattern. They are therefore synchronous with each other.

The new object, the audio word object, and the current procedural memory pattern are in resonance. The forward synchronous projections of these two declarative objects to higher levels of declarative association cortex are now going to create an association pattern that is going to bind these two objects together. The association of these objects is built into association cortex in higher levels of association cortex.

The forward projections from these two declarative objects resemble waves caused by two rocks simultaneously dropped into water. Those wave patterns will intermix through declarative association cortex like two radio antenna interference patterns and where the peaks of both waves add, the association areas will fire. Those association areas that fire in response to receiving these two waves build the cortical pattern that is the declarative association pattern between the new object and the audio word object. The actual association pattern is dependant on the patterns of the two objects and where their forward projections are additive. This is a completely new and unique pattern that associates the two object patterns within your one image deep declarative storage projection screen.

This new associative declarative object projects to the frontal cortex and causes its mirror image pattern to light up in your procedural screen and project feedback to the associative object. At this point the object, audio word object and associative object all have you attention and are in resonance with their mirror images in your procedural storage screen. The object pattern, word pattern, and associative pattern are now all part of the object's expanded declarative pattern. You now know what the object is called. All associations are created in this manner.

Either the object pattern or word pattern lighting up in your declarative projection screen will cause the entire declarative object to light up. Hearing the word causes the word pattern to light up and forward projection of the word pattern causes the association object to become active. Forward projections of the word object and association pattern causes the procedural mirror image of the entire declarative image to light up. Procedural feedback coupled with association object feedback causes the entire declarative object, including the object pattern, to light up. Both the word and object now have your attention.

The highest levels of declarative cortex are extremely large and are reciprocally interconnected with equally large prefrontal procedural cortical areas. These final pattern detection areas are unique to humans. No other animal comes close to the number of levels of cascaded declarative cortex and the size of these final association areas. It is these final association levels that support the pattern detection that enables the associations that makes human intelligence possible.

The highest levels of declarative association memory receive input from virtually all areas of lower declarative memory. An association object pattern at these highest levels is reciprocally connected with all of the possible input patterns that can cause it to be recognized. Each object addition to a declarative memory pattern creates a new association pattern object and the overall declarative pattern grows to include both new objects. When the highest levels of a mirrored object pattern light up in your declarative and procedural cortex, feedback to lower levels of declarative memory causes all of the input patterns that collectively define the object to light up. The feedback path from the highest levels causes pattern recognition in all representations of the object. All declarative associations and abstractions are supported in this manner.

At some point in your past you saw an apple for the first time. The visual apple pattern in your temporal cortex was projected to your prefrontal cortex and became synchronized with the current procedural memory. You asked what the object was and then heard the word "apple" spoken. The audio pattern of the word apple then was projected as a separate object pattern to your frontal cortex and it also became synchronized with the current procedural memory. The two separate objects, the visual apple pattern and the audio apple pattern, became synchronous with each other. The synchronous projection of the two object patterns of apple caused higher levels of declarative cortex to fire in recognition of the summation of both apple pattern inputs. The declarative pattern of apple became more complex to include both representations of apple.

You store declarative memory patterns very quickly. One learning event is sufficient to create the association between the visual apple object pattern and the spoken word apple object pattern. You eat the apple and its taste and smell patterns cause the association apple object to expand to include them. In this fashion, all of the inputs that can evoke the apple object pattern become a part of the overall declarative pattern for apple. Any of these inputs will cause the large complex apple pattern to light up in your declarative cerebral projection screen.

An apple is a fruit. Fruit is a word that describes a class of objects. The declarative pattern for the word fruit is associated with all of the declarative objects in non-frontal cortex that comprise that class of objects. Pattern recognition of any one of these objects will cause the declarative pattern for the word fruit to light up. The declarative pattern for the word fruit is a part of the declarative pattern of all of the objects that it is associated with. Fruit is a word object only. The only inputs that cause the fruit object to be recognized are direct language sensory input of the word fruit or pattern recognition of any of the class of declarative objects that fruit describes.

Fruit is a secondary abstraction, a rule. Language allows word objects to be created that are rules that can be associated with other objects through declarative learning. Rules are themselves declarative objects and therefore rules can be created and applied to them in a recursive manner. For instance, food is a word pattern that is associated with the declarative object fruit. Only humans have the cascaded levels of cortex that allow the creation of an enormous set of rules that apply to virtually all objects and space and allow humans to learn to understand their world.

The building of associations and abstractions between declarative objects enables human learning. For two separate declarative objects to become associated, their patterns must first both become a part of the declarative input to the frontal cortex and basal ganglia that is utilized to select a procedural memory pattern. The nucleus reticularis then causes both the current procedural memory and the two separate declarative objects to become synchronous. The now synchronous input from the two objects allows them to become part of a larger declarative pattern at the highest levels of declarative memory.

The capability to build complex multi level patterns that include multiple representations of virtually everything you have experienced is what enables human intelligence. It is what allows you to build the abstractions and associations that are the goals of education. An object pattern in human declarative memory includes the object's highest-level pattern, all of the object's sensory input patterns, and a vast set of rules and definitions that apply to the object. Any input that causes the object's highest-level pattern to light up causes all of these associated patterns to light up. This enables language, mathematics, history, geography, science and virtually all of human learning.

Procedural Learning – Storing the Patterns That Drive Your Behavior

Human behavior is caused by the interaction of two learning systems. The learning of procedural memories reflects your entire experience, all of the combinations of declarative input and limbic states you have ever encountered. The cortical memory pattern occupying the procedural projection screen drives actual behavior. The basal ganglia support behavioral learning, the modification of your response to declarative and procedural input based on the expectation of a future outcome. Learning in the basal ganglia supports the selection of an overall procedural memory pattern to drive behavior. Procedural learning is a combination of building cortical procedural memories and basal ganglia behavior patterns.

Procedural memory is a mirror image of declarative memory. In humans, declarative memory contains an amazing amount of information. The environmental space you occupy, all of the objects in that space, the declarative patterns for each of those objects that includes all associations, abstractions and recursive rule sets that apply to each of those objects, the objects caused to become active due to procedural input, what you are thinking, and all of the associations, abstractions and recursive rule sets that apply to each of those objects. The procedural memory pattern driving your behavior contains a mirror image of all of these patterns. As you look around, you know a great deal about every object you see. You understand your world and how it works.

The construction of procedural memory is a shared process between the declarative and procedural systems. Declarative and procedural memories are constructed at the same time. The building of a declarative memory requires recognition and feedback from procedural memory. A new procedural memory pattern is created that mirrors each and every new declarative memory pattern. The sum total of your experience stored in your frontal cortex is continuously modified every day to incorporate the unique declarative input that drove your behavior that day.

Behavioral learning is a function of the basal ganglia. Your basal ganglia have produced a behavioral response procedural pattern to literally every environmental and limbic state that has ever lit up your procedural projection screen. You literally never encounter exactly the same declarative memory coupled with the same limbic state twice. The closest matching procedural memory lights up and is accepted or modified by the basal ganglia. Basal ganglia modification will occur if the behavior specified by the procedural pattern previously caused an increase or decrease in dopamine input to the basal ganglia. That change in dopamine input is heavily influenced by limbic input to the basal ganglia. If the basal ganglia accept the procedural pattern without modification, the learned behavior specified by the procedural pattern will be carried out.

Procedural learning involves the creation and modification of procedural memories coupled with the modification of the basal ganglia's response to those procedural memory patterns. Human behavior results from procedural patterns driving three separate processes. Procedural motor patterns combined with cerebellar motor patterns drive skeletal muscle fibers, mirror images of declarative patterns projected to the declarative system cause those declarative patterns to become active, and procedural patterns projected to the amygdala and hypothalamus exert influence over the limbic system.

In our projection system analogy, the patterns that represent objects and space are still pictures. The pattern that represents an apple is static. A procedural memory pattern is part of a movie. The procedural memory patterns are frames of the movie. All learned behavior is stored in the frontal cortex as a sequence of procedural memory patterns.

All learning involves both the procedural and declarative systems. Declarative learning is building understanding. Procedural learning is building behavior to cope with experience. Learning is a process that allows you to store ever more complex abstractions of your world and develop ever more complex behaviors to cope with that world.

Short-Term Memory

Based on our architectural model, short-term memory is really not memory at all but a function of the basal ganglia. The basal ganglia controls frontal thalamocortical loops and via that control holds procedural memories active until the purpose of the behavior is accomplished. When you receive a telephone number from the operator, each number pattern lights up in your declarative cortex as you hear the number. The mirror image of each number lights up in your frontal cortex as the active procedural memory is altered slightly to include each new declarative number pattern. The return projection reinforces the declarative number pattern and the pattern becomes synchronized with the procedural memory. As long as your basal ganglia holds the procedural memory pattern active and therefore synchronized with the seven number declarative patterns, you can recall the number. If your basal ganglia allow the frontal cortex to move on to something else, the declarative patterns for the numbers turn off and the telephone number is gone. Short-term memory is the ability of frontal cortical procedural memory under the control of the basal ganglia to hold disparate declarative patterns active and therefore connected for a short time. The number of distinct declarative memory patterns that you can hold on in this manner is limited to around seven.

Behavior - Thinking

The human brain has evolved to the point where humans have the ability to ignore environmental input and think. This ability to think is a uniquely human capability. Thinking is implemented in prefrontal cortex as procedural memory drives declarative storage. Procedural feedback can elicit declarative patterns of real or imagined events. The declarative patterns evoked through procedural memory feedback are often abstractions of real world objects in the form of language. Humans first talked to themselves and then learned to disconnect speech motor output in order to have a private conversation. That private conversation now dominates your waking life.

Thinking involves the same behavioral movie that drives language. The language generated is for internal use only. The procedural memory patterns that drive the behavior of thinking are constantly in use and are well learned. Through them you have access to and control over your entire declarative storage. When you're thinking behavioral movie causes a word declarative object to become active, the entire collective declarative object that includes all input representations of the object and all rules associated with the object also become active. The overall thinking cortical pattern contains your procedural memory pattern synchronous with the declarative patterns currently enabled by it. Thinking declarative patterns always have your attention. Under basal ganglia control the thinking behavioral movie flows as we access new words and declarative patterns.

Thinking is not necessarily driven by real world events. Thinking allows you to manipulate declarative patterns and build more powerful patterns that are representations completely unique to yourself. The declarative patterns thus constructed are just as valid as patterns learned from primary environmental input. The ability of the human brain to internally create and manipulate declarative memory is what allows imagination and creativity.

Language

Human language may be the driving force that caused intelligence rather than intelligence enabling language. The level of communication complexity supported by human discourse is amazing. When someone speaks to you, each of their words causes the declarative pattern for the word, what it represents and all abstractions and rules associated with the word to light up in your declarative storage projection screen. Your procedural screen projects back the mirror image of this entire declarative pattern as you attend to what is being said.

The speaker is driving your declarative and procedural storage projection screens, your entire memory pattern. The patterns that result from listening are equivalent to the patterns that are generating by thinking the same sequence of words. The speaker is driving your thoughts. Exactly the same effect is happening now as you read these words.

The learning of language requires mastering the abstraction of all objects and the related system of rules concerning those abstractions. Words are abstractions that are manipulated by a set of rules. In order to be accomplished in language you must build within your declarative cortex a pyramid of rule based abstractions and build within your frontal cortex a multitude of solution behaviors. The quantity of hierarchical neural pattern recognitions required for competency in language is so large that you must begin their construction in childhood in order to become proficient.

You have words that are abstractions of virtually everything you perceive in your environment. A child of four typically has a vocabulary of over three thousand words and that number increases throughout life. Language involves heavy use of object based declarative memory and is mostly a left brained activity.

The portion of declarative cortex most associated with language is Wernicke's area. Wernicke's area is located in the temporal lobe in close association with the primary auditory cortex. Your first exposure and learning of words is through auditory input and you develop a vast vocabulary before you learn to read. The audio pattern recognitions that represent words are supported by Wernicke's area association cortex in close proximity to sensory auditory cortex. When you begin to learn to read, you link the new written word visual object patterns with the auditory object patterns in Wernicke's area. Your well-developed auditory language object recognition area, Wernicke's area, remains your primary language association area.

The portion of procedural cortex most associated with language is Broca's area located in prefrontal cortex just forward of premotor cortex that controls the mouth and tongue. This is the area that controls your actual speech. The behavior of following a set of rules for the construction of language is also well learned long before you learn to read. Again, since your first learning of language is auditory, the cortical area that produces auditory language is your first and dominant language output center.

Around the age of six, you begin to add another symbolic representation to virtually every declarative object in your declarative cortex, the written word. Visual declarative patterns of words are synchronized with the objects they name and made a part of a larger declarative memory pattern for those objects as you learn to read. You begin this learning process by reading aloud to bind the visual word with the auditory pattern representation. With training, the ability to write and read achieves equal status with auditory communication in your ability to abstract your world.

Your first exposure to language is auditory. Your mastery of language occurs while auditory representations are your only language modality. The complex set of rules that govern the proper construction and utilization of words are learned as a behavior that produces auditory speech. As you add abstractions to your language capabilities in the form of the written word, these input associations are tied into the declarative memories you have already formed. The cortex that supports auditory input and output remains your primary language cortical centers.

Consciousness
(Exactly Where in Your Brain Are You?)

The brain architecture presented allows us to speculate about some of the neural aspects of what we term consciousness. We are going to examine the arguments that the conscious you resides in one of the three main neural subsystems, the declarative system, the procedural system or the limbic system. We will make the case for each of these subsystems as the neural embodiment of you.

We begin with evidence that you reside within your declarative memory system. You continuously build from primary sensory input a neural representation of the environmental space that surrounds your body. That representation includes all of the objects within that space. One of those objects is you. In fact, the object that is you is always present in your neural representation of your environment. If you did not have an active neural object representation of you, there would be a blank hole in your neural representation of your environmental space. The activation and recognition of the declarative object you is the minimum neural capability required for what we term consciousness.

In humans, the declarative memory pattern that represents you has reached gigantic proportions and dominates your declarative memory cortex. The brightest object within your declarative cerebral cortex is the abstraction of you. The construction of this self-object neural pattern starts perhaps before birth or shortly thereafter and is altered with every waking moment of your life. All of your sensations, emotions and experiences have impact on your declarative object representation of yourself. The declarative object pattern you dominates all other patterns stored in your cerebral cortex. Its representation is so strong that it remains active at all times.

Your procedural memory pattern remains constantly synchronized with your declarative memory pattern of yourself, you are at all times aware of yourself, you have your constant attention. This constant synchronization of your prefrontal procedural memory pattern to the declarative cerebral neural pattern that represents you is consciousness.

The declarative memory pattern that is you is so well developed that virtually everything causes it to be recognized as you interpret the world as it relates to you. The patterns of the space you occupy and the objects you interact with are made synchronous with your current procedural memory and the dominant declarative pattern that represents you.

The declarative pattern that represents you is always a bright pattern on your cortical projection screen. The other patterns caused by sensing your environment or driven by prefrontal input coexist with this self-declarative pattern. Your prefrontal cortex and basal ganglia are constantly pattern detecting declarative input patterns that include your pattern. Your pattern is a part of every procedural memory selection.

You interpret the world as it relates to you. From a functional point of view, the declarative object that is you has taken control of your cortex, conscious control.

Now for the argument that you reside within your procedural memory system. The declarative memory system is really just a storage facility. Holding the patterns for objects, space, and you that are available for the operation of the procedural system. The procedural system must reflect all declarative patterns in order for those patterns to remain active. This includes all space, all objects and all associations between objects. It controls all declarative learning. The procedural system is in complete control of the declarative memory system.

The procedural system contains your entire behavior repertoire and continuously controls which of those possible behaviors is active. Your behavior defines you. The procedural system is what controls thinking, creativity and intelligence. It also provides conscious control over your limbic system. It has the ability to control how you feel. From a functional point of view, the procedural memory system has taken conscious control of your declarative system and your limbic system.

Finally, the argument that your limbic system is the real you. The limbic system is the oldest of the three neural subsystems that can lay claim to consciousness. The limbic system was the first neural system to exercise overall control of the brain and it has never really given up that control. It controls how you feel. How you feel controls everything else.

Declarative memories are only stored if the limbic system causes them to be stored through its control of the hippocampus. All learning of objects, space, abstractions and associations is caused by the limbic system. If your limbic system does not care or is uninterested in something, you do not store it. How you feel drives your declarative system.

Procedural memories are also only stored if the limbic system causes them to be stored in the same manner. In the case of procedural memories, direct limbic input is a part of the pattern detection that comprises procedural memories. How you feel directly influences which procedural memories light up. Limbic input to the striosomes within the basal ganglia determines the level of dopamine to be administered as a result of the current active behavior. In this manner the limbic system controls the learning of selection of actual behavior. How you feel controls how you behave. Your limbic system controls all learning in both the declarative and procedural systems.

Your limbic system literally keeps you alive. It has been providing overall control for brains ever since its evolution hundreds of millions of years ago. There has never been an evolutionary need for it to relinquish control. How you feel drives everything you learn and everything you do. Your expanded human declarative and procedural systems are great additions to brain architecture, but your limbic system remains firmly in charge. So where are you inside your brain? All three systems can lay a legitimate claim to contain you or at least a portion of you. The discussion of what consciousness is and where it resides will continue.

Conclusion

Your brain is a machine. There are over 6.5 billion people on this planet and every one of their brains is exactly like yours, the same components wired in exactly the same way, interconnected in the same way and performing the same functions. The embryonic growth of the brain, the pruning of synapses and neurons, the storage of procedural and declarative patterns, literally everything the brain does, is exactly the same in all people. There is no difference between their brains and yours. Your brain is a machine, you are not.

The storing of declarative and procedural patterns that record your life and knowledge and define your behavior, enabled by the machinery of your brain, are completely unique to you. Your existence is separated in time and space from everyone else. The complexity of your experience and diversity of your learned behavior are different from everyone else. You are not a machine. The cerebral recorded movie of your life is one of a kind.

This chapter contains a great amount of speculation. The brain architecture postulated is a model based on current research. The human brain functions described are based on the architectural model. The description of the human brain presented here is based on analysis of currently available knowledge. Based on that knowledge, this explanation seems to best fit the available data.

When we began this analysis, the prospect of constructing a human brain from components that only performed pattern detection seemed improbable. The brain architecture proposed satisfies that constraint. The human brain is implemented by large parallel collections of neurons with each neuron performing only pattern detection.

Writing this book has been an incredible learning experience. I will never forget it.

Index

77, 82-86, 102, 107, 118,
121, 126, 133, 135, 137,
158-163, 166, 169, 185,
226-127, 231

Stellate Neurons, 50, 69-72,
144, 175, 192, 199, 204-
207, 222, 248

Striatum, 76-80, 108, 128,
214-220, 223, 231,

Substantia Negra, 42, 74-78,
80, 122, 171, 173, 214-
215, 218-219, 225

Superior Colliculus, 42, 139-
143, 149-150

T

Temporal Cortex, 19, 32, 96,
108, 143, 148, 151, 184,
197, 201, 214, 245, 251,
264

Thalamocortical Loop, 54,
56, 58, 65, 69, 79, 188,
190-193, 196, 199, 203-
225, 229, 231, 243, 246-
249, 254, 268

Thalamus, 13, 18, 20-21, 23,
25-26, 29-34, 39-40, 42,
51-56, 58, 65-73, 76-77,
79-80, 82, 85, 96, 99,
106-108, 119, 122-123,
126-127, 132-133, 135,
139, 143-144, 146, 149,
152, 155, 157-158, 162,
165-166, 169, 173, 180,

182, 184-188, 193, 196,
200-204, 209-210, 213,
215-221, 225, 229, 231,
233, 238, 240-243, 246,
248, 250, 252-255, 258-
259, 267

Tracts, 4, 23, 40, 84, 121,
133, 161-162, 166, 172

V

Visual, 19-20, 24-31, 41-42,
55-56, 61, 63, 66-68, 81,
90, 92-93, 96, 99, 107-
108, 111, 122-125, 132,
138-157, 169, 185, 190-
191, 193, 197, 200, 208,
210-212, 225, 229-230,
246, 250-251, 261, 264,
270-271